Superconductivity: A Very Short Introduction

VERY SHORT INTRODUCTIONS are for anyone wanting a stimulating and accessible way into a new subject. They are written by experts, and have been published in more than 25 languages worldwide.

The series began in 1995, and now represents a wide variety of topics in history, philosophy, religion, science, and the humanities. The VSI library now contains 200 volumes—a Very Short Introduction to everything from ancient Egypt and Indian philosophy to conceptual art and cosmology—and will continue to grow to a library of around 300 titles.

Very Short Introductions available now:

For more information visit our web site
www.oup.co.uk/general/vsi/

Stephen Blundell

SUPERCONDUCTIVITY
A Very Short Introduction

OXFORD
UNIVERSITY PRESS

OXFORD
UNIVERSITY PRESS

Great Clarendon Street, Oxford OX2 6DP

Oxford University Press is a department of the University of Oxford.
It furthers the University's objective of excellence in research, scholarship,
and education by publishing worldwide in

Oxford New York

Auckland Cape Town Dar es Salaam Hong Kong Karachi
Kuala Lumpur Madrid Melbourne Mexico City Nairobi
New Delhi Shanghai Taipei Toronto

With offices in

Argentina Austria Brazil Chile Czech Republic France Greece
Guatemala Hungary Italy Japan Poland Portugal Singapore
South Korea Switzerland Thailand Turkey Ukraine Vietnam

Oxford is a registered trade mark of Oxford University Press
in the UK and in certain other countries

Published in the United States
by Oxford University Press Inc., New York

British Library Cataloguing in Publication Data

Data available

Library of Congress Cataloging in Publication Data

Data available

Typeset by SPI Publisher Services, Pondicherry, India
Printed in Great Britain by
Ashford Colour Press Ltd, Gosport, Hampshire

ISBN 978-0-19-954090-7

1 3 5 7 9 10 8 6 4 2

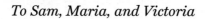

To Sam, Maria, and Victoria

Contents

Contents

Acknowledgements

I would like to record my thanks to Katherine Blundell, Tom Lancaster, and Latha Menon for many helpful comments on drafts of this book.

List of Illustrations

Superconductivity

Chapter 1
What is superconductivity?

'Is it a bird? Is it a plane? No, it's superman!': Clark Kent's alter ego
was conceived in the 1920s and the name 'superman' owed more
than a little to Nietzsche's concept of an *Übermensch*, an 'over-
man' who would rule over inferior beings. The superhero thus
began life as more morally ambiguous than the champion of good
and defender of the helpless that he later became. Nevertheless, he
was created at a time when the prefix *super* was beginning to be
used to enhance all manner of concepts. Until the 20th century,
super was usually found in rather technical words: superannuation,
supererogatory, superposition, supervise, and superintendent.
But the era that launched the caped crusader also provided
supermarket, supertanker, and superstar. Later came
supercomputer, superglue, supergrass, supermodel, superpower,
supersize, and the superstore. More recently, the MRSA bacterium
has become a hospital superbug.

When Kamerlingh Onnes, working in his Leiden laboratory
in 1911, discovered a strange phenomenon affecting the
properties of mercury at very low temperature, he christened it
suprageleider. When translated from Dutch into English, it
became *supraconductivity*, but this rapidly mutated into
superconductivity. Though Onnes' naming predated Superman
by more than a decade, the spirit was rather similar: just as the
comic-book hero could defy gravity in a way no mere man could,

the superconductor could defy the usual laws of electricity in a manner never before achieved by any known material. Superconductors were not just better than ordinary conductors of electricity, they were of a completely different order, as strange and mysterious as a visitor from the planet Krypton wearing underpants over his trousers.

Materials can be classified in terms of how well they conduct electricity. Metals, like copper and gold, are good conductors of electricity and are used for making wires. The electrons in a metal are able to easily travel around. Many (though not all) plastics and rubbers tend to be poor conductors of electricity, and are classified as insulators: the electrons in an insulator are fixed in place and are not able to travel around easily. Insulators can be used to make the shielding that is wrapped around wires so that you don't electrocute yourself when you touch the outside of a cable. In between the extremes of metals and insulators are semiconductors (a 19th-century term expressing the halfway-house nature of these beasts) and examples include silicon and germanium. Semiconductors behave very much like insulators, but they can be persuaded to conduct electricity by adding impurities and they find uses in transistors and computer chips. Onnes' discovery though was something else entirely.

To appreciate how bizarre superconductivity is, imagine making a coil of superconducting wire and somehow passing an electrical current around it. Never mind for a moment how you would do this, we will get to that later. What you would find is that the electrical current would keep going round and round the coil *forever*. Once started, the current keeps on going. Batteries are not included for this experiment because you wouldn't need them. You can retreat to a safe distance and watch the extraordinary sight of a current going round and round, all by itself, with no power source driving it. This looks like perpetual motion, a concept which down through the ages is normally associated with fools and charlatans. But this is no conjuring trick: it has been done. People have set electrical currents going round and round superconducting coils for years and even decades, and with no source of power the currents keep going all by themselves for as long as anyone can be bothered to do the experiment.

Superconductivity is a phenomenon that simply did not exist before the 20th century; there was no hint or sign and barely any suspicion that such a thing might be possible. Yet, as we shall see, the seeds of the discovery of superconductivity were planted early in the 19th century. Once superconductivity was discovered, it would take nearly half a century for a satisfactory theory explaining it to be developed and the succeeding half a century would throw up surprising experimental puzzles which would show that our understanding of the effect is far from complete. Nevertheless, these coils of superconducting wire which carry electrical currents all by themselves are used daily in MRI (magnetic resonance imaging) scanners in hospitals throughout the world and in the Maglev trains in Japan. Superconductivity actually works, and earns its living every day of the year.

To see how superconductivity is fundamentally different from normal behaviour, consider the following. When a wire carries an electrical current it gets hot, an effect known as Joule heating, named in honour of the 19th-century Lancashire brewer-turned-scientist James Prescott Joule, who discovered it. The effect is normally small, but in a fuse the heat generated by a large unwanted current causes the fuse wire to melt and break the circuit. Fuse-breaking is a dramatic effect but all wires would act like fuses if they were thin enough; wires have to be made sufficiently thick so that the heating due to the current which they carry is small enough so that they don't melt and break. Why do wires heat up when they carry current? To understand this, think of the carriers of electrical charge in a metal, the electrons, as a swarm of angry bees, each one zipping around in some apparently random direction. Driving a current is like trying to gently waft the swarm in a particular direction by subjecting them to a breeze, so that even though each bee is rushing back and forth at great speed, the swarm as a whole drifts along with the breeze. However, the bees keep bumping into things, slamming into tree branches and hedges, and even though each emerges unscathed, these collisions dissipate energy and serve to heat up, ever so slightly, the objects with which the bees collide. It is this transfer of energy from the cloud of electrons into the surrounding atoms that means that a lot of the Earth's precious energy is wasted in heating up the miles of power cables that crisscross our cities and connect them to power stations. These collisions are responsible for heating up the elements in our kettles and the wires in our toasters, applications in which the heating effect is put to good use. But in power cables, the Joule heating is just a waste of energy.

However, in superconductors the Joule heating is entirely absent. It is as if the friction has been turned off and the crowd of angry bees waft gently through the garden without bumping into anything. A superconductor can carry a current with no electrical resistance and so, somewhat counterintuitively, you can cause a current to go round and round a superconducting coil for ever and ever without supplying any power! If superconductors could be made to operate at room temperature, it could revolutionize the way we supply electricity to peoples' homes and have many important consequences for our technology. Kamerlingh Onnes realized this soon after the discovery of superconductivity and envisaged making superconducting wires and winding them into coils. These could then act as large electromagnets, producing large magnetic fields without needing any source of energy to drive them. Today this dream has become a reality: the large magnets used in hospital MRI scanners are made of coils of superconducting wire.

However, there is a fly in the ointment. For superconductivity to occur, the material has to be cooled to a very low temperature.

In fact, as we will describe in the next chapter, it was only by the development of methods to cool matter to extremely low temperatures that superconductivity was discovered at all. We now know that it might be possible to get superconductivity to work at room temperature, but although we are more than half way to achieving that goal, we still do not know how to do it.

Temperature is a quantity that is measured by certain scales (see Figure 1): the Fahrenheit scale dating back to 1724 and still used in the United States, and the Celsius scale which was developed about 20 years later and is widely used in Europe. Both scales are based on setting certain values to the fixed points defined by the freezing point and boiling point of water. Of course, other substances boil at different temperatures (see Figure 1) and these temperatures can be measured on these scales. Daniel Gabriel Fahrenheit's original scale used the temperature measured under a human armpit as one of its three fixed points, while Anders Celsius' original scale had two fixed points but was curiously arranged upside down (zero degrees was set as the boiling point of water, with 100 degrees as the freezing point of water, so that hotter temperatures meant lower numbers, a feature corrected by the Swedish botanist and taxonomist Carl Linnaeus).

In any case, during the 19th century it was realized that temperature quantifies the degree to which a system can vibrate and fluctuate by exchanging energy with its environment; warmer molecules 'jiggle around' more than cold ones. At a certain temperature, found to be $-273.15°C$ ($-459.67°F$) and known as *absolute zero*, all vibrations cease. Lord Kelvin defined a new temperature scale, one which bears his name, as simply the Celsius scale shifted by 273.15 degrees; hence absolute zero becomes zero Kelvin (abbreviated to K) and the freezing point of water becomes 273.15K. Although Celsius and Fahrenheit are convenient scales for discussing the weather, we will use Kelvin in this book as the numbers are more suitable for discussing the low temperatures

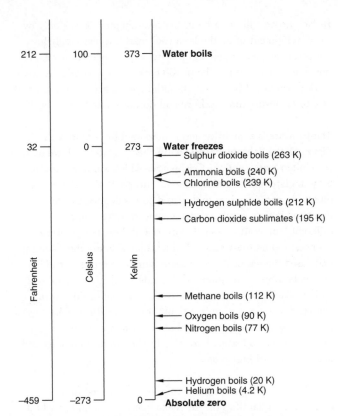

Superconductivity

Fahrenheit	Celsius	Kelvin	
212 —	100 —	373 —	**Water boils**
32 —	0 —	273 —	**Water freezes**
			◄— Sulphur dioxide boils (263 K)
			◄— Ammonia boils (240 K)
			◄— Chlorine boils (239 K)
			◄— Hydrogen sulphide boils (212 K)
			◄— Carbon dioxide sublimates (195 K)
			◄— Methane boils (112 K)
			◄— Oxygen boils (90 K)
			◄— Nitrogen boils (77 K)
			◄— Hydrogen boils (20 K)
			◄— Helium boils (4.2 K)
−459 —	−273 —	0 —	**Absolute zero**

1. **The Fahrenheit, Celsius, and Kelvin temperature scales**

at which superconductivity occurs (and it is now the Kelvin scale which forms the absolute standard from which the Celsius and Fahrenheit scales are derived). It was the quest for low temperatures in the 19th century which paved the way for the discovery of superconductivity in the 20th century.

Chapter 2
The quest for low temperatures

Liquefying gases

Though superconductivity was not discovered until 1911, the origins of the discovery can be traced back at least to the early 19th century and the work of Michael Faraday in the Royal Institution in London. At age 20, Faraday, an apprentice bookbinder from a poor family, had managed to secure a job at the Royal Institution as scientific assistant to the eminent chemist Sir Humphrey Davy, mainly on the strength of presenting to Davy a bound version of the notes Faraday had taken at some of Davy's public lectures. Though Davy's wife persisted in treating Faraday as a servant from the lower classes, and Davy himself was later to block Faraday's progress in the scientific establishment as he realized his own eminence was about to be eclipsed, Faraday was forever grateful and devoted himself to a life of constant hard work in the laboratory. Though he is best remembered for his work in electromagnetism, optics, and electrochemistry, it was his accidental discovery of how to make liquid chlorine that was to be of such importance in the road to superconductors.

Chlorine had been discovered in 1774 by Carl Scheele and was thought to contain oxygen because of its strong oxidizing properties; it was thus named *oxymuriatic acid*, muriatic acid being what we now call hydrogen chloride (HCl). Davy had

2. Michael Faraday

triumphantly shown that oxymuriatic acid did not react with hot
carbon and thus contained no oxygen and pronounced it an
element, naming it chlorine after its greenish-yellow colour.
Chlorine gas was later to make a somewhat mixed contribution to
human happiness following its use in disinfecting swimming pools
and killing soldiers in the trenches of the First World War.
However, it was a compound of chlorine which was to lead to a
discovery that is very significant for our story.

In 1811, Davy had showed that the crystals obtained by passing chlorine gas through a nearly freezing, dilute solution of calcium chloride were a compound of chlorine and water: chlorine hydrate ($Cl_2 \cdot H_2O$). In the winter of 1823, at Davy's suggestion, Faraday performed what turned out to be some crucial experiments on chlorine hydrate. 'I took advantage of the late cold weather to procure crystals of this substance', he described in his report. Faraday placed the crystals 'in a sealed glass tube, the upper end of which was then hermetically closed'. He heated the tube and noted the formation of an coloured oily liquid on subsequent cooling. The best results were performed using a bent tube; he heated one end with the chlorine hydrate in it and allowed the oily liquid to condense in the cold end which was submerged in crushed ice. By performing experiments on this liquid, Faraday realized that what he had made was liquid chlorine.

We now understand that chlorine gas needs to be cooled to about $-34°C$ to liquefy, and this is colder than any winter Faraday was likely to encounter in London. However, the high pressure produced by decomposing the chlorine hydrate, which occurs on heating it in the sealed tube, was enough to raise the boiling point of chlorine to the temperature in Faraday's laboratory. The same effect, only in reverse, is responsible for the poor quality of tea that one can brew at the top of mountains; the reduced air pressure lowers the boiling point of water so that less flavour is extracted from the tea leaves.

Faraday had showed that a substance previously known only in the gaseous form could be turned into a liquid. He now wondered whether he could perform the same trick with other gases. Through further experimentation, it was found that this technique of producing high pressures in a sealed tube allowed one to liquefy other gases, including ammonia (NH_3), hydrogen sulphide (H_2S), nitrogen dioxide (NO_2), sulphur dioxide (SO_2), and carbon dioxide (CO_2). Carbon dioxide misses out the liquid phase at normal

9

pressures. Solid CO_2 turns straight into gaseous CO_2 when you heat it (a process called sublimation) and is used as 'dry ice' (particularly in cheesy music videos). Faraday's work was pioneering but though he was the first person to liquefy a chemical element, the compound ammonia had in fact first been liquefied using pressure back in 1787 by the Dutch chemist Martinus van Marum. However, despite intensive effort, there were certain gases (including hydrogen, nitrogen, and oxygen) that Faraday was unable to liquefy using this technique and for this reason he called them *permanent gases*.

In the 1860s, the Belfast-born physicist Thomas Andrews studied the liquefaction of gases in great detail. He formulated the conditions under which liquefaction could occur, connecting these conditions to the gas laws which relate the pressure, temperature, and volume of the gas. It was then realized that the only reason that the so-called permanent gases had stubbornly resisted liquefaction was simply that the pressures available in a Faraday-style experiment were insufficient to raise their boiling temperatures up to room temperature (see Figure 1). A more cunning approach was needed and it came by accident.

A sudden release

Louis Paul Cailletet was the son of a metallurgist and had set up a laboratory at his father's iron foundry. Cailletet had extended Andrews' work and had made careful measurements of how the properties of gases deviated from the laws proposed by the Dutch physicist Johannes van der Waals. In the 1870s, using his souped-up version of Faraday's tried and trusted method of applying high pressure, Cailletet began his attempt to turn gases into liquids at room temperature, identifying acetylene (C_2H_2) as a likely candidate. It was expected that a pressure of about 60 atmospheres was needed to produce the desired effect, but during the pressurization his apparatus sprang a leak and the compressed gas escaped. Cailletet had been watching carefully and noticed that as the gas escaped through the leak, a faint mist had formed, only to rapidly disappear. He initially suspected this must be water vapour and that his sample of acetylene had been impure, but repetition of the experiment with a more carefully purified sample of acetylene produced the same result. He realized that the sudden release of pressure in the gas had cooled it and resulted in a temporary condensation of liquid. Very high pressure wasn't necessary to liquefy gases; you could do it by suddenly releasing the pressure!

Cailletet had the gumption to realize that this was a breakthrough and quickly set about trying to liquefy something more interesting than acetylene. He started with oxygen because he could make a reasonable quantity in a pure state, pressurized it to 300 atmospheres and cooled his glass apparatus to $-29°C$ with evaporated sulphur dioxide. Suddenly releasing the pressure produced a mist of condensing droplets of liquid oxygen. He reported his results to the Academy of Sciences in Paris in December 1877, only to find that at the same time they had received a report of a similar discovery by the Swiss chemist Raoul-Pierre Pictet based in Geneva. Pictet, who had been motivated to liquefy gases in order to produce artificial ice for food preservation, had achieved the same result but by a quite different 'cascade' method which consists of liquefying gas which has been precooled in the liquid of another substance that has in turn been produced from gas itself precooled by another liquid. In this way, a succession of harder-to-liquefy gases can be produced.

Two methods were thus available to liquefy gases, and it seemed that the quest to liquefy the permanent gases was nearly at an end, Cailletet furthering the quest by liquefying nitrogen and carbon monoxide. Many copies of Cailletet's apparatus were manufactured in Paris (since Cailletet was very open about all his experimental details and admirably keen to see his results repeated and extended). One set was bought by a Polish scientist called Zygmunt Florenty Wróblewski, who was taking up the chair of the Faculty of Physics at the Jagiellonian University in Kraków. There

Wróblewski began an initially fruitful partnership with a colleague in the Chemistry Department, Karol Olszewski, and, by making some modifications to Cailletet's equipment, they were able to do more than simply produce a fine mist of liquid droplets of oxygen: in March 1883, they produced liquid oxygen quietly boiling away by itself in a test tube! Two weeks later, they repeated the trick with liquid nitrogen and Kraków instantly became the world-leading centre of low-temperature physics. Unfortunately, Wróblewski and Olszewski had a serious falling out and their professional relationship broke up after a further six months. Thereafter they worked independently in their own departments, despite working on precisely the same project, that of attempting to liquefy hydrogen. Toiling late one night in his laboratory in 1888, Wróblewski upset a kerosene lamp on his desk and was so badly burned he died soon afterwards. Olszewski continued to work on low-temperature problems, developing an improved Pictet-style apparatus.

The principle of cooling by rapid expansion had been established much earlier, in 1852, by James Prescott Joule, together with the Belfast-born mathematical physicist William Thomson, later to be known as Lord Kelvin. Their effect is known either as the *Joule–Thomson effect* or, reflecting Thomson's later elevation, as the *Joule–Kelvin effect*. It works because, as a gas expands, the average distance between molecules increases and this alters the effect of the weak intermolecular attractive forces. It turns out that the Joule–Thomson effect only leads to cooling if the gas is already at a lowish temperature but, this complication notwithstanding, the effect is hugely important for liquefying gases.

One method of getting gases to expand was by allowing high pressure gas to squirt out of a fine nozzle or constriction into a region of low pressure. This would cool the gas, allowing it to liquefy, and any cold gas remaining could be recompressed and forced around a circuit and back into the high pressure vessel. In this way, a steady flow process could be produced and a gas liquefier could be constructed that could chug away nicely by itself, steadily producing precious drops of extremely cold liquids. This feat was perfected by Carl Paul Gottfried von Linde who, in the early 1870s, had set up an engineering laboratory in Munich (Rudolf Diesel, inventor of the Diesel engine, was one of the students). His work on refrigeration led to the development of the Linde gas liquefier and his first commercial refrigeration system was patented and installed in 1873. He founded 'Linde's Ice Machine Company' in 1879, which is now the Linde group and at the time of writing the world's largest industrial gas company with annual sales of well over 10 billion euros.

By the mid-1870s, the most important known gases had been liquefied, apart from one: hydrogen. This had stubbornly refused to liquefy, though the Kraków scientists had seen some fine droplets, but it was not clear if these had been impurities. However, there was one unknown gas that was to prove more important to liquefy and this was the first element to be discovered beyond the Earth.

A new element on the Sun

In 1868, the French astronomer Pierre Janssen was in India, studying the spectrum of light coming from the Sun's chromosphere (a thin layer of the Sun's atmosphere) during a total solar eclipse. He noticed a bright yellow line with a wavelength of 587.49 nanometres which was initially assumed to be due to sodium. Later that year, the same line was observed by the English astronomer Norman Lockyer, later to be the first editor of the journal *Nature*. Lockyer concluded that this spectral line must be due to a new element, unknown on Earth but present on the Sun. He and Edward Frankland (a Professor of Chemistry at the Royal Institution) named the element *helium* from the Greek word for the Sun (*helios*).

In 1895, the Scottish chemist William Ramsay isolated helium in the laboratory of University College London by treating the mineral cleveite with mineral acids. Ramsay was looking for argon but, after separating nitrogen and oxygen from the gas liberated by sulfuric acid, noticed a bright-yellow line that matched the line observed in the spectrum of the Sun. Lockyer and the physicist William Crookes were able to confirm the identity of the gas as helium. We now know that helium is trapped in various minerals because of radioactivity: alpha particles are helium nuclei and so helium is being continually produced inside the Earth due to radioactive decay processes. At the same time as Ramsay's discovery, helium was independently isolated from the very same mineral by chemists in Sweden who managed to collect a sufficient quantity of the gas to accurately determine its atomic weight. Ramsay scooped the Nobel Prize in 1904 'in recognition of his services in the discovery of the inert gaseous elements in air, and his determination of their place in the periodic system', reflecting not only his discovery of helium but also that of the other noble gases: argon, neon, krypton, and xenon.

Liquefying the lightest element

Now that helium had been discovered, the race to liquefy both hydrogen and helium was on and one person determined to be first in that race was Sir James Dewar. Dewar had been educated in Edinburgh University and, after a spell at Cambridge, had in 1877 become Fullerian Professor of Chemistry at the Royal Institution, the chair first held by Faraday. The following year he obtained a Cailletet apparatus from Paris and within a few months was demonstrating droplets of liquid oxygen to the great and the good at one of the Royal Institution's Friday Evening Discourses. Years of work were necessary to catch up with the Polish scientists, but in 1886 Dewar succeeded in producing solid oxygen.

3. Sir James Dewar

Dewar, always keen that research should be publically viewable in one of his lecture demonstrations, wanted his cryogenic liquids to be boiling quietly in a test tube and one problem was that glass vessels with very cold liquids inside them tend to frost up. This makes the cold liquids invisible and, worse, is a sign that heat is seeping into them from the outside world. What was needed was a container for keeping the cool liquids nice and cold but still allowing them to be visible. Dewar went to work on the problem and eventually he was ready to demonstrate the result of hours of careful design, patient thought, and meticulous glass-blowing. In January 1893, to an audience at a Friday Evening Discourse, Dewar unveiled his famous double-walled container which became known as a vacuum flask or a 'dewar'. They are now often known by their trade name 'thermos flask', though the modern versions that contain your hot coffee are no longer transparent, and frequently (though unaccountably) decorated with tartan. The region between the walls is evacuated to minimize heat loss through conduction and convection, and heat loss due to radiation can be minimized by applying a reflective coating (silvering) to the inner walls.

Dewar's quest was to produce the lowest temperature possible, even to attempt to achieve absolute zero. He deduced that to do this he would have to first liquefy the lightest known element: hydrogen. Since hydrogen was so light, it would be the hardest to liquefy; cool that down and you have the coldest possible liquid. Any substance in contact with liquid hydrogen would itself become liquid or solid.

Further improvements to Dewar's liquefaction techniques continued through the 1890s and finally, on 10 May 1898, Dewar produced about twenty cubic centimetres (about five teaspoonfuls) of liquid hydrogen, a result which was announced at the Royal Society two days later. However, it was not entirely clear how cold the liquid hydrogen was since Dewar's electrical thermometer gave a nonsensical reading and had clearly failed to function at these low temperatures. He tried his best with a gas thermometer instead and deduced (accurately as it turned out) that his liquid hydrogen was about twenty degrees above absolute zero (the modern value is 20.28K or $-252.87°$C). He also thought he had liquefied helium at the same time, but it turned out that what he had taken to be condensed helium was in fact an impurity. The following year, Dewar managed to produce further cooling and turned his liquid

4. Sir William Ramsay and Sir Joseph Norman Lockyer

hydrogen solid (at just below 14K or −259°C), which he established did not conduct electricity.

However, Dewar soon realized that not only had he not liquefied helium, but that helium obstinately remained gaseous, even when cooled to the lowest temperature he had achieved so far, a temperature which was low enough to solidify hydrogen. Liquefying hydrogen was not the final step on the road to absolute zero after all: the real prize was to liquefy helium.

Dewar versus Ramsay

Dewar had several big advantages in the race to liquefy helium. He had been the first person to make liquid hydrogen, and he had the first person to isolate helium, William Ramsay, working within walking distance of his own lab. Unfortunately, Dewar and Ramsay had fallen out and were quite unable to work together. The problems between the two had started when Ramsay had pointed out, at Dewar's moment of triumph in announcing to the Royal Society that hydrogen had been liquefied, that Karol Olszewski had already done it. As Dewar later wrote, in 1895 following his reporting of preliminary results, 'Professor William Ramsay made an announcement of a sensational character, which amounted to stating that my hydrogen results had been not only anticipated but bettered.' However, 'Professor Olszewski published no confirmations of the experiments detailed by Professor Ramsay in scientific journals of date immediately preceding my paper or during the following years 1896, 1897 or up to May, 1898.' At this point, Dewar announced his final triumph, but the 'moment the announcement was made by me to the Royal Society in May, 1898 that, after years of labour, hydrogen had at last been obtained as a static liquid, Professor Ramsay repeated the story to the Royal Society that Olszewski had anticipated my results.'

In fact, Ramsay had got his wires crossed and Olszewski readily admitted that he had been unsuccessful and that Dewar had beaten him to it. Ramsay was forced to read a new letter from Olszewski at the following meeting of the Royal Society communicating this. Dewar wrote in the journal *Science*, somewhat grumpily:

> This oral communication of the contents of the new Olszewski letter (of which it is to be regretted there is no record in the published proceedings of the Royal Society) is the only kind of retraction Professor Ramsay has thought fit to make of his published misstatements of facts. No satisfactory explanation has yet been given by Professor Ramsay of his twice repeated categorical statements made before scientific bodies of the results of experiments which, in fact, had never been made by their alleged author.

It is not surprising that Ramsay and Dewar were barely on speaking terms. Ramsay (with his knowledge of helium) and

Dewar (with his expertise in low temperatures) could have made an invincible team and been the first to liquefy helium. But it was not to be. As described in the following chapter, low-temperature physics in the first decades of the 20th century was going to be dominated by a laboratory in neither London nor in Kraków, but in Leiden.

Chapter 3
The discovery of superconductivity

Through measurement to knowledge

The winner of the race to make liquid helium was Heike
Kamerlingh Onnes and because of this, he was also to be the
discoverer of superconductivity. Onnes was born in Groningen in
1853 and studied under Robert Bunsen (now famous for his Bunsen
burner) and Gustav Kirchhoff (now famous for formulating various
laws in circuit theory and thermal physics) in Heidelberg. He
returned to Groningen in 1873 to pursue his doctoral work on the
influence of the Earth's rotation on the motion of a pendulum, but
towards the end of his doctoral work he became acquainted with a
professor at the University of Amsterdam, Johannes Diderik van
der Waals. The influence that van der Waals' thinking was to have
on Onnes was enormous. Van der Waals had been on a quest to
provide a coherent description of the properties of gases. He
realized that the theory of the ideal gas, developed by Robert Boyle
and others in the 17th century, was woefully inadequate to describe
the properties of real gases. The most glaring omission of the
standard theory was that it failed to predict that gases would liquefy
if cooled sufficiently; this came about because it completely ignored
the intermolecular forces that exist between molecules in a gas, and
if these are absent then nothing will induce a gas to condense into
liquid. In 1873, van der Waals had succeeded in providing a law
which included these forces and successfully related the

5. Kamerlingh Onnes (seated, left) and Johannes van der Waals

temperature at which a gas would liquefy to the strength of the intermolecular forces. In 1880 he published his famous law of corresponding states which provided a single equation that should describe the behaviour of all real gases. As an experimentalist, Onnes was fascinated by these theoretical developments and realized that in order to test these predictions it was important to measure, as accurately as possible, the behaviour of real substances at very low temperatures.

In 1881, Onnes was himself appointed to a chair in Leiden and it was here that he set about to build his world-famous low-temperature physics laboratory. His central goal was to provide, in the extreme conditions of very low temperature, accurate and reliable measurements that could test the latest theories to the limit; the laboratory motto was therefore the poetic *Door meten tot weten* ('Through measurement to knowledge'). Onnes was one of the first people to really understand that advances in this field depended critically on having first-rate technicians, expert glass-blowers, and skilled craftsmen to build and maintain the delicate equipment and provide support for the technically demanding experiments; it was not enough to have a lone eccentric individual pottering around in a ramshackle laboratory. Onnes therefore brought a much-needed level of professionalism into experimental science and the production of liquid gases, an attitude which was singularly lacking in the laboratories of his British competitors.

Onnes founded a Society for the Promotion of Training of Instrument Makers which was crucial for building up the necessary skilled workforce. He expected much from his team and it was said of him that he 'ruled over the minds of his assistants as the wind urges on the clouds.' At his funeral in 1926, his technicians had to follow the cortège in black coats and top hats from the city to a nearby village churchyard. Outside the city, the horse-drawn hearse went at a brisk pace and the technicians arrived at the graveyard sweating and panting. One of them is reported to have said 'Just like the old man; even when he is dead he keeps you running.'

Onnes' character was however suited to getting the most out of his highly trained team. Though he could be demanding, he combined this with a kindly nature and extreme politeness, naturally instilling respect and loyalty. As Dewar was prickly, disputatious, and secretive, so Onnes was friendly, benevolent, and open. Onnes welcomed visitors into his laboratory and was more than ready to discuss his work, listen to suggestions, and to collaborate. Dewar was the exact opposite.

In the 1890s, Onnes was perfecting his cascade process for the production of liquid gases. Work was held up when, in 1896, Leiden town council got wind of Onnes' possession of large amounts of compressed hydrogen. This brought back dark memories of a catastrophe that had affected the city 89 years earlier when, during Napoleon's occupation, an ammunition ship had exploded in a canal in the centre of the city. The resulting hullabaloo shut down Onnes' laboratory for two years, despite van der Waals being appointed to sit on the council's investigating commission and Dewar generously sending a helpful letter pleading for Onnes' research to be allowed to continue.

Onnes only succeeded in liquefying hydrogen in 1906, eight years after Dewar had achieved the same feat, but Onnes' apparatus produced much larger quantities and his apparatus was much more reliable. Onnes was playing the long game and this was to bear fruit when he attempted to liquefy helium. In this quest, he was also able to use a family advantage: his younger brother was director of the Office of Commercial Intelligence in Amsterdam and in 1905 was able to procure large quantities of monazite sand from North Carolina; helium gas could be extracted from the mineral monazite (a few cubic centimetres from each gram of sand) and after three years of work Onnes had over 300 litres of helium gas at his disposal. By this time, he was also able to make more than 1,000 litres of liquid air in his laboratory, easily enough to run his cascade apparatus. He was now ready to attempt to make liquid helium.

On 10 July 1908, the experiment began to run, helium gas flowed through the circuit and the temperature fell. However, after fourteen hours of work, there was no sign of liquid helium and the temperature stopped falling and seemed to be stuck resolutely at 4.2K. It was suggested that this might be because liquid had already formed but was hard to see, and this in fact turned out to be the case. Onnes adjusted the lighting of the vessel, illuminating it from below, and suddenly it was possible to perceive the liquid–gas interface. Onnes wrote: 'It was a wonderful sight when the liquid, which looked almost unreal, was seen for the first time ... Its surface stood sharply against the vessel like the edge of a knife.'

Liquid helium had been made in the laboratory, and in quantity: sixty cubic centimetres filled the vessel, enough to fill a tea cup! Onnes concluded: 'Faraday's problem as to whether all gases can be liquefied has now been solved step by step in the sense of van der Waals' words "matter will always show attraction" and thus a fundamental problem has been removed.' The Nobel Prize for Physics in 1913 was awarded to Kamerlingh Onnes 'for his investigations on the properties of matter at low temperatures which led, *inter alia*, to the production of liquid helium'. In his Nobel lecture, he recalled: 'How happy I was to be able to show condensed helium to my distinguished friend van der Waals, whose theory had guided me to the end of my work on the liquefaction of gases.'

Resistance is useless?

Now that liquid helium could be made, Onnes could investigate its properties. For a start, he tried reducing the pressure above the surface of liquid helium and succeeded in cooling the liquid to 1K. He tried to make helium become a solid but failed to do so (we now know this quest was hopeless and that helium will only solidify at high pressure). He improved his liquefier so that it produced a litre of helium every three to four hours in 1908 and up to nearly a couple of litres an hour a decade later. He also managed to find a way to store liquid helium in a helium cryostat so that experiments on materials at low temperature could be performed.

Onnes decided to turn his new-found experimental technique to an outstanding scientific problem of the day. What would happen to the resistance of a metal as it was cooled to absolute zero? It was already well known that the resistance of a metal fell as you cooled it. We now understand this as being due to the reduction of vibrations of the atoms in a solid that accompany cooling. The atoms in a solid wobble around like a vibrating jelly, but there is less jiggling around at colder temperatures. The electrons move through the solid when you try to pass an electrical current, but

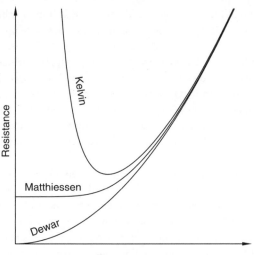

Temperature

6. The low-temperature resistance of metals according to three popular theories at the turn of the 20th century. But which one would agree with experiment?

electrons can be deflected when they interact with vibrating atoms, and this deflection is called 'scattering'. At low temperature, fewer vibrations mean less scattering of electrons and the resistance falls. In much the same way, your passage across a rope bridge spanning a yawning chasm is considerably aided if it doesn't bounce around too much. But what happens as the temperature falls to absolute zero?

There were three possible theories which were in vogue in the first decade of the 20th century (see the diagram in Figure 6). Dewar was convinced that the resistance would drop inexorably to zero as the temperature fell. Lord Kelvin insisted that the electrons themselves would start to freeze, impeding further flow and causing the resistance to rise. Much earlier, Matthiessen had claimed that the resistance would flatten out at some low but non-zero value. As Onnes had access to the world's lowest temperatures in his laboratory, he was uniquely placed to settle the dispute. His instinct told him that Kelvin was probably right, but he knew that experiment would be the final arbiter. Just as his work on liquid helium was motivated by a desire to establish the governing equation determining the properties of a gas (following the ideals of van der Waals) so his work on conductivity was motivated by a desire to establish a similar governing equation for electrons.

Dewar's early work on this problem had been on samples of silver and gold cooled to only about 16K and there were already indications that Matthiessen might be right. One problem was that impurities were always present and these gave a route for scattering the electrons and hence increasing the resistance in a manner that no amount of cooling could bypass. This necessitated using exceptionally pure samples to determine the intrinsic behaviour of the metal.

It was clear that a limiting factor was the impurity content of the metal and so Onnes chose to focus on mercury, a liquid metal, which could be repeatedly distilled to make it as pure as possible. To make wires of mercury, his technician filed very fine U-shape glass capillaries and then carefully froze them. The capillaries had electrodes at either end so it was possible to pass an electrical current through them and follow the resistance at various temperatures. These samples were cooled using Onnes' newly discovered liquid helium, allowing him to reach much lower temperatures than Dewar could obtain. The experiments were not easy however as the resistance of the mercury was very low, and Onnes and his team had to proceed carefully and methodically.

The experiments in 1911 showed that when the mercury sample was slowly cooled below the boiling point of liquid helium (4.2K), the resistance of mercury disappeared suddenly. Onnes put this down to a short circuit appearing in his experimental apparatus, but repeated attempts failed to remove the apparent experimental artefact. Light only dawned following a mistake: on one trial, the temperature in the cryostat was being controlled by a junior assistant who was responsible for adjusting a valve to maintain the vapour pressure of helium. The assistant nodded off and the temperature began to rise, and suddenly the resistance in the mercury reappeared. Onnes realized that the state of zero resistance was not an experimental artefact but a real state of the mercury which set in once mercury was cooled below a certain *critical temperature*. He had discovered superconductivity.

Data from Onnes' experiments are shown in Figure 7 and show that while gold behaves in the manner predicted by Matthiessen (its resistance falls and levels off at a constant value due to the effect of impurities) the resistance in his sample of mercury, while larger at high temperatures, plummets much more dramatically

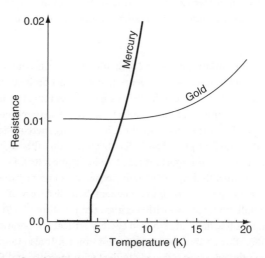

7. **The results of Onnes' experiments showing the measured resistance of gold, which doesn't superconduct, and mercury, which does**

and below about 4K performs what he began to realize was the incredible superconductivity vanishing trick.

First steps

Of course determining a zero resistance state is hard to do. Onnes realized that all he had established was that the resistance of his sample of mercury suddenly dropped to an unmeasurably low value. After all, if you put a grain of salt on your bathroom weighing scales you will not notice any change, but you do not from this conclude that the grain of salt has no weight, merely that its weight is unmeasurably small (at least unmeasurable with your bathroom scales). Nevertheless, the ability of an electrical current to pass round a superconducting circuit without observable diminution, something that Onnes was soon able to demonstrate, was convincing evidence that the superconducting state was something very remarkable. Within a few months of his initial discovery, Onnes was able to show that the resistance of mercury decreases at the critical temperature by a factor of at least ten billion.

Somewhat infuriatingly, Onnes' team discovered subsequently that adding small impurities to mercury had no effect on its transition temperature; the superconductivity appeared as robust as in pure mercury. This implied that the effect was intrinsic to mercury. It also implied that the time and effort in distilling mercury to purify it in the first place had in fact been in vain (although this observation was an important clue to what came later). For a while it seemed that the effect only occurred in mercury due to some strange peculiarity of that element. The discovery of superconductivity in tin (with a critical temperature of 3.7K) in 1912, and a few days after in lead (with a critical temperature over 6K), showed that mercury was not unique. It also thankfully removed the necessity of having to deal with mercury and Onnes was able to perform some further explorations of this new phenomenon with samples of tin and lead, much easier substances to work with. He was able to construct a circuit in which he could start a current flowing in a superconductor using a battery, but then disconnect the battery from the circuit and observe the current continuing to flow.

Onnes' lab in Leiden had access to large supplies of liquid helium and so for a long time had a monopoly on work in superconductivity. It was therefore some time before any other laboratory could catch up. Onnes' work had pointed the way to how future laboratories needed to be equipped, and he wrote in his Nobel lecture in 1913: 'In the future I see all over the Leiden laboratory measurements being made in cryostats, to which liquid helium is transported just as the other liquid gases now are, and in which this gas also, one might say, will be as freely available as water.' Onnes thus accurately foresaw the subsequent growth of low-temperature physics but, at least in his lifetime, the remarkable phenomenon of superconductivity which he had discovered remained entirely inexplicable.

Chapter 4
Expulsion

By the early 1930s, a number of experimental facts about superconductivity had been discovered and, as more experiments were performed, the number of superconducting elements was slowly increasing (see Figure 8). It was known that in these particular materials, the resistance would decrease to zero when cooled down below a critical temperature. The zero resistance state reached at low temperature could be destroyed if the material was subjected to a sufficiently large magnetic field, or if the current passing through the superconductor exceeded a critical amount (though Francis Silsbee, of the National Bureau of Standards in Washington had showed in 1916 that the critical current and critical magnetic field were two sides of the same coin). Those materials included the metals mercury, tin, lead and gallium, none of them particularly good conductors of electricity, but emphatically not copper, silver, and gold, which are excellent conductors of electricity. This in itself was an important piece of the jigsaw but no-one could understood its significance at the time. Infuriatingly, experiments showed that the critical magnetic field which would destroy superconductivity was rather small. This was a blow to Kamerlingh Onnes. He had hoped that superconductors could be used to make wires for magnet coils.

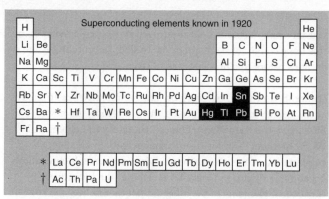

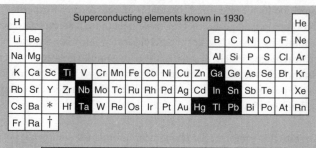

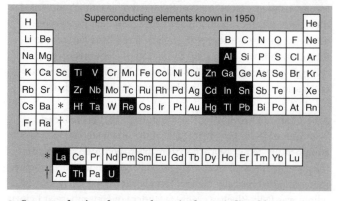

8. Superconducting elements shown in the periodic table, as known in 1920, 1930, and 1950. A modern version is shown in Figure 33

If even small magnetic fields destroyed superconductivity, then this hope could not be realized. As we shall see, it took several decades before this problem was solved.

Onnes died in 1926, but work on superconductivity continued at the Leiden laboratory. In 1931, W. J. de Haas, the new director of the Kamerlingh Onnes Laboratory in Leiden, and W. H. Keesom, discovered superconductivity in an alloy, a combination of metallic elements. Many superconducting alloys began to be discovered, and it seemed that the individual constituent elements in the alloy did not themselves need to superconduct. For example, a sample of four percent bismuth dissolved in gold performed the trick, even though neither gold nor bismuth by themselves are superconductors. Surprisingly, the critical magnetic field in some of the alloy samples greatly exceeded that which was seen in elements. Work on alloys was to dominate much research in the next twenty or thirty years and lead to many technological applications.

Despite this progress, the phenomenon remained completely without a satisfactory explanation. Sir J. J. Thomson, the English physicist who in 1899 had discovered the electron in the Cavendish Laboratory in Cambridge, came up with a complicated and utterly erroneous theory based on the alignment of atomic dipoles inside a metal. Frederick Lindemann at Oxford, who had successfully devised a theory of the melting solids, fared little better with a formulation involving a rigid formation of electrons drifting through a lattice of ions. The field was fair game for ingenious fabrications but what was lacking was a really good idea.

Albert Einstein, reviewing the situation in 1922, concluded that 'with our wide-ranging ignorance of the quantum mechanics of composite systems, we are far from able to compose a theory out of these vague ideas. We can only rely on experiment.' Nevertheless, many fine theoretically inclined minds continued to try, one of whom was Felix Bloch, a graduate student of Heisenberg at Leipzig who in 1928 had gone to work with Wolfgang Pauli in Zurich and who would achieve later recognition following his contributions to solid state physics. Bloch found his attempts to formulate a satisfactory theory of superconductivity were doomed to failure; every avenue he tried seemed to end in a brick wall. Bloch concluded that 'the only theorem about superconductivity that can be proved is that any theory of superconductivity is refutable'. His equally facetious second theorem was: 'Superconductivity is impossible.' Before we can describe how this impasse was broken, it is time to step back a bit and think about Ohm's law.

Ohm's law

Ohm's law is named after Georg Simon Ohm (1789–1854), a German physicist and high school teacher who published it in 1827 in a textbook about the properties of electricity. In fact, Ohm's law was probably first discovered in 1781 by the eccentric and reclusive English physicist Henry Cavendish (1731–1810). Cavendish was extremely wealthy but devoted himself to his scientific investigations which included determining the composition of the air, determining the density of the Earth (by measuring the gravitational attraction of two 350 pound lead spheres using a torsion balance, producing an answer extremely close to the modern value) and performing experiments into electricity. Cavendish had no instruments capable of measuring electric current (the galvanometer was not invented until ten years after his death) so managed the feat by passing the current through his own body and assessing his level of pain. Though physical pain held no terror for him, the company of other human beings certainly did, particularly women, and he apparently had a back staircase added to his house in order to avoid encountering his housekeeper with whom he communicated by means of leaving handwritten notes (his order for dinner was almost invariably 'a leg of mutton'). Cavendish, a tall thin man with a high-pitched squeaky voice, remained solitary and secretive and much of his work lay unpublished. His later relatives endowed the Cavendish

33

Laboratory in Cambridge (which opened in 1874) and its first professor, James Clerk Maxwell, gained access to Cavendish's papers and found out how many discoveries credited to others had in fact first been made by Cavendish.

Ohm's law states that the voltage applied to a metal is equal to the product of the current through it and the metal's electrical resistance. One can think of this by analogy to the flow of water down a pipe: the voltage driving the electrical current is analogous to the pressure of water driving a flow of water; how much pressure is needed depends on the resistance of the pipe which will depend in turn on the length and width of the pipe. Ohm's law works for many different metals and gives a good description of the way in which electricity is conducted through them.

Perfect conductors?

If superconductivity was all about perfect conductivity, then it was quickly realized that there were consequences of such behaviour. One of these followed from an insight made by Michael Faraday: a changing magnetic field induces a voltage in a conductor. This effect is at the root of how a bicycle dynamo or an electric turbine works. As the bicycle wheel rotates, driven by your furious pedalling, magnets are rotated inside the dynamo and the resulting varying magnetic field induces a voltage around a coil of wire, thus driving an electric current into your bicycle lights. This all works because the coil of wire is a conductor of electricity, but what if it were a superconductor? A superconducting wire could support no voltage around it because were it to do so then, by Ohm's law, an infinite current would have to flow. Faraday's insight implied that the magnetic field inside a superconductor could never change. Once a material became superconducting, as you cooled it through its transition temperature, it was argued that the magnetic field that happened to be there 'at birth', so to speak, would remain with it until you warmed it up through its transition temperature. This implied that a superconductor would 'trap' any magnetic field present when you cooled it down and keep that trapped magnetic field until you warmed it up again. Experiments seemed to show that this trapping of magnetic field actually occurred. However, these experiments were not correct, as became clear following the discovery of what has become known (somewhat unfairly to Robert Ochsenfeld) as the Meissner effect.

In the early 1930s, the techniques of low-temperature research were beginning to be developed in other laboratories around the world and Leiden was slowly losing its monopoly on research at helium temperatures. The Meissner effect was discovered in Berlin, one of the places in which the new low-temperature physics research had got going. In 1933, Walther Meissner and Robert Ochsenfeld were performing an experiment to look in detail at what happens to the magnetic field near a superconductor when it is cooled down through its transition temperature. Contrary to expectation, they managed to show that the magnetic field was not *trapped in* the superconductor but appeared to be *expelled from* it. The magnetic fields that were quite content to pass through the material at high temperatures are rudely evicted as soon as the temperature is low enough for superconductivity to occur. The magnetic field lines now have to pass around the superconductor, doing a circuitous detour as if sensing some kind of invisible 'NO ENTRY' sign. It turned out that earlier results that seemed to show trapping were due to magnetic field being trapped in impure, non-superconducting parts of the sample.

The Meissner effect, this forcible expulsion of magnetic fields from a superconductor, is responsible for the incredible ability of superconductors to levitate above magnets (or indeed for magnets to levitate above superconductors), as shown in Figure 9. The magnetic field is expelled from the superconductor due to electrical currents which run across its surface. These are known as screening currents, because they act to screen the interior of the superconductor from the externally applied magnetic field. These currents also produce a magnetic field exterior to the

9. Levitation of a superconductor

superconductor which acts in opposition to the magnetic field you started with and repels the magnet. This repulsive force balances the force of gravity and allows the superconductor to hover eerily in mid-air. We will see later that this leads to a number of important applications.

The discovery of the Meissner effect was clearly significant. But what did Meissner and Ochsenfeld's results mean?

Understanding the Meissner effect

One of the first people to appreciate the significance of the Meissner effect was Cornelis Jacobus Gorter who had studied physics in Leiden and finished his doctoral work with de Haas in 1932. Gorter had moved to Haarlem and started to think about the implications of the Meissner effect. He concluded that superconductors were more than just 'perfect conductors' and that the observation of magnetic field expulsion indicated that a superconductor could only really exist in the absence of a magnetic field. If you placed a superconductor in a magnetic field, currents would form on its exterior in order to force the magnetic field lines out of its interior. A superconductor in the presence of a magnetic field is a bit like Superman faced with a kryptonite sandwich; just as Superman's powers drain away when he comes into contact with that mythical green mineral, so a superconductor will not tolerate magnetic fields passing through it. A fundamental part of its very existence seems to be defined by its absence of magnetic field. This was a crucial idea, but seemed very unfamiliar to many physicists; Wolfgang Pauli was unconvinced about this idea and even Meissner was sceptical about Gorter's conclusions.

Gorter also understood that Meissner's observation meant that superconductivity was a well-defined thermodynamic state. Up until that point, it was felt that since superconductors sometimes would trap a magnetic field inside them and sometimes wouldn't, their low-temperature state was not very well defined because it depended on the sample's history. In other words, it depended on precisely how it was cooled. This meant that superconductivity was not what is known as an 'equilibrium state of matter'. To understand this point, think of the transition of water between its different phases: solid (ice), liquid (water), and gas (steam). If I give you a glass of cold water, you have no way of knowing whether that water has been condensed from steam and then cooled, or whether I melted a large ice cube and allowed the water to warm up a bit. Changes of phase are entirely reversible and once a substance has come to equilibrium with its surroundings, there is no memory of its past history. To describe the water in the glass, I just have to detail its properties at this current moment such as the temperature of the water in the glass right now. However, if superconductors are not 'equilibrium states of matter', then you can't describe them by their current properties but need to know their complete history; how they got to their current temperature. A consequence of this was that, in describing superconductivity, one could not utilize all the sophisticated apparatus of equilibrium thermodynamics, as developed in the 19th century, which had been so successful in describing the properties of matter and which beautifully describes phenomena such as the melting of ice and the condensing of steam.

Now that it was understood that the trapping of magnetic fields was an experimental artefact, it was also seen that the superconducting state was well defined after all. Superconductors are in fact 'equilibrium states of matter' and their properties do not depend on the history of the sample. This understanding left the way clear for the whole panoply of thermodynamical techniques to be used. In 1934, Gorter, together with Hendrik Casimir, plunged right in and formulated a 'two-fluid' model of superconductors: they supposed a superconductor to contain both normal electrons (which don't superconduct) and superelectrons (which do) and the two species were to co-exist. The idea was that as you cooled the superconductor, the fraction of superelectrons increased at the expense of the normal electrons. As you warmed it back up, the superelectron fraction decreased, falling to zero at the transition temperature. This theory made some predictions and could be made to agree with experimental data, but the theory was a little *ad hoc* and it seemed that something was still missing. The necessary insight came to a pair of remarkable brothers who had been driven out of Germany because of the political events of the 1930s.

The London brothers

Fritz London was born in Breslau in 1900. After a brief dalliance with philosophy, he switched to physics and became immersed in the heady intellectual atmosphere of the 1920s that followed from the development of the new quantum theory. The new ideas that came from the architects of quantum mechanics, Bohr, Heisenberg, Schrödinger, and their co-workers, provided wonderful new possibilities for explaining many different phenomena that had previously been understood only at an empirical level. Bohr's theory of the hydrogen atom had been one of the first triumphs of the new physics and the race was on to take quantum mechanics into chemical problems. Fritz London seized this challenge and obtained appointments in theoretical physics, first with Paul Ewald in Stuttgart and then with Arnold Sommerfeld in Munich. London really made his name by working out a theory to explain the hydrogen molecule and thus founding a new discipline: quantum chemistry. This project was done with Walter Heitler while both were in Zurich in 1927, supposedly working with Schrödinger. However, the great man was not really very interested in working with anybody else and very much left Heitler and London to get on with it. The Heitler–London theory was quickly recognized as a major achievement and is taught in university courses today.

Fritz London next used quantum mechanics to explain weak intermolecular forces (today called 'London dispersion forces') and thus established himself as a young physicist with a growing reputation and track record. However, like many young scientists both then and since, London had not achieved a permanent job but had been funded on a series of short-term appointments, each time spending a few years in a particular university city but then having to move on.

Fritz's younger brother, Heinz, was born in Bonn in 1907 and also became a physicist, studying with the low-temperature expert Franz Simon at Breslau. Simon had come from Berlin where his research group had had to avoid research in superconductivity because this was Meissner's patch; in Breslau, this inhibition was no longer valid and Heinz London started to work in superconductivity, first of all experimentally, but then turning his attention to theoretical aspects.

In the early 1930s the dark cloud of fascism was rising over Germany. The attendant anti-Semitism had profound consequences for many physicists with Jewish ancestry, and that included both London brothers. A possible way out was provided from an unlikely source: Frederick Lindemann, half-German himself but now ensconced in Oxford, decided to do what he could to provide a safe haven for refugee scientists in Oxford. His motives were not entirely altruistic; Oxford's physics department was a bit of an intellectual backwater and, at the time, greatly outclassed by Cambridge's Cavendish Laboratory. This was a way to effect an instantaneous invigoration of Oxford's intellectual firepower in

10. Fritz London and Heinz London

physics, and in 1933 he persuaded the chemical company ICI to come up with funds to support this endeavour.

Both Erwin Schrödinger and Albert Einstein were lured to Oxford, although Einstein quickly moved on to Princeton. Another import was Franz Simon (in England he became Francis Simon, and later Sir Francis Simon) who had won the Iron Cross for his service in the First World War. However, as a Jew, even such an obvious demonstration of patriotism did not leave him immune from the ire of the new regime in Germany. Simon arrived, together with various experts in his group: Heinz London, Kurt Mendelssohn, and Nicholas Kurti. Lindemann also wanted a theoretician and admired Fritz London as a no-nonsense, practical sort of person who was able to work on down-to-earth problems, and so both London brothers ended up in Oxford. Fritz London and his wife moved into a house in Hill Top Road, Oxford, and Heinz stayed with them, giving the brothers an opportunity to talk and work together about superconductivity. Their joint work was to provide the biggest breakthrough yet in understanding the field.

Updating Ohm's law

In the 1920s and 1930s, the quantum revolution was in full swing and it was realized that Ohm's law needed updating and that it was necessary to formulate a quantum theory of electrical conduction. This could be built on the work of Paul Drude, a student of Heinrich Hertz, who had begun his research at a time when Maxwell's equations of electromagnetism were being thought about in Germany. Drude had focused on applying James Clerk Maxwell's theory of electromagnetism to describe the optical properties of materials and found a generalization of Ohm's law. His model considered an electron gas which drifts in the direction of an applied voltage and assumed that the motion of electrons was damped by collisions with ions.

This was however a purely classical theory, and so the baton passed to Arnold Sommerfeld, a German theoretical physicist, who tried to extend Drude's results (Drude had sadly committed suicide in 1906). Sommerfeld had an extraordinary legacy in physics with six of his students going on to win Nobel Prizes. Those who worked under him and are mentioned in this book include Fröhlich, Heisenberg, Heitler, London, and Pauli. Einstein once said to Sommerfeld: 'What I especially admire about you is that you have, as it were, pounded out of the soil such a large number of young talents.' In 1927, Sommerfeld used results developed by Enrico Fermi and Paul Dirac, describing the statistical properties of electrons in detail (and known as Fermi–Dirac statistics), to formulate a theory for the electrical transport in a metal based on Drude's assumptions. Though Drude's model had been classical, Sommerfeld's use of Fermi–Dirac statistics put some quantum polish on it. The Drude–Sommerfeld model gave a pretty good description of the electrical properties of ordinary metals. But it could not describe superconductors.

The London equations

The London brothers were trying to figure out how to make something like Ohm's law, or the Drude–Sommerfeld model, work for superconductors. Ohm's law itself wasn't going to help since superconductors allow a current to flow all by itself, with no voltage driving it. However, they concluded that the Meissner effect gave them the clue as to how it worked.

The Meissner effect is now known to be a more fundamental property of superconductivity even than the zero-resistance state. The expulsion of magnetic fields was seen by Fritz London to be an indicator that the electrons in a superconductor were behaving in a very curious way. Normally, the electrons which carry the current in a wire are completely independent of each other, in much the same way as the voices in a room of people all having different conversations are independent. The electrons in a superconducting wire are not independent of each other but act together, almost as if they were a single entity, rather like the voices in a choir singing in unison. This understanding was the key to explaining the expulsion of magnetic fields.

The Meissner effect shows how a superconductor responds to a magnetic field. Therefore, the London brothers wrote down an equation which connects the current in a superconductor, not with an applied voltage, but with the magnetic field. Specifically, they found a way to connect the current in the superconductor with something called the magnetic vector potential, a concept introduced by James Clerk Maxwell to solve various problems in electromagnetism. Their resulting equation could be used to explain the Meissner effect because the current flowing along the surface of a superconductor screens the interior from magnetic fields, producing an expulsion of magnetic field. It shows that the current in a superconductor is not driven by an electric field (as for a normal metal). Instead, it just exists, all by itself, wrapped up in its own magnetic field so to speak.

The London brothers had started with the realization that Maxwell's theory of electromagnetism simply didn't contain the physics of superconductivity and that something else was needed. Their conclusion that the superconducting current was associated with a magnetic field was the key ingredient. Their equation allowed them to show that magnetic fields are excluded from a superconductor, as Meissner and Oschenfeld had found, but that the fields could penetrate a very short distance into the surface of a superconductor. This distance was related to the mass, charge and number of the superconducting carriers and emerged naturally from their equation. This distance is called the London penetration depth, and very soon it was a quantity which had been measured in all known superconductors.

But what does it all mean?

It is one thing to come up with a revolutionary new equation, but why does it work? As explained already, at the time in which the Londons were working, the whole of theoretical physics was undergoing something of a revolution due to the development of the new theory of quantum mechanics and the London brothers were in the first generation of physicists to have been born into it. Many of the old certainties which had driven the breathtaking advances of physics in the 19th century were now in doubt. The most basic of concepts in physics, such as momentum, position, and energy, which had been the rocks on which the whole edifice had been built, now acquired a shimmering quantum halo. Do you really know where the electron that carries a charge is, or how fast it is going? What truth is actually out there?

Schrödinger had showed that particles of matter, like electrons, actually behaved like waves and could be described by what was

43

called a 'wavefunction', given the symbol ψ. If you look at ψ at a point in space and at a particular time, it is a bit like looking at a boat on the ocean waves. The wave has an amplitude (how far the boat goes up and down as the waves pass) and a phase (at what point of the cycle the boat is on, whether at a maximum or a minimum or somewhere in between). Another theoretical physicist, Max Born, proposed in 1926 that the probability of finding a particle in a particular place was really only to do with the amplitude of the wave and not the phase (in fact, he said the crucial quantity was the so-called 'modulus square of the wavefunction', given by $|\psi|^2$, but that only carries information about the amplitude). This might make you think that the phase of a wavefunction has no role to play. However, the phase becomes important when two waves combine. The interference between electron waves passing through two slits produces the famous diffraction pattern observed in the 'two-slit experiment', the focus of many of the early discussions about quantum mechanics. When two waves have the same phase, they can combine constructively. When the two waves have opposite phase, they combine destructively (see Figure 11).

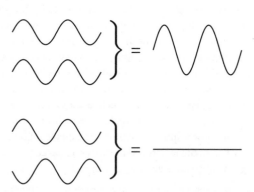

11. Two waves combine (top) with the same phase, resulting in constructive interference, or (bottom) with opposite phase, resulting in destructive interference

Although Arnold Sommerfeld's theories had used quantum mechanics, his focus had been on the probability of the electron distribution. The phases of the individual wavefunctions had been ignored. However, that appeared to be very sensible since it was reasoned that they would all be different and all the possible interference effects would surely cancel out. It turned out that although this is normally true, in a superconductor the phases of the wavefunctions do not all randomly cancel out. This is because they all have the *same value*. In a superconductor, the phases of the individual electron wavefunctions are all locked together. This locking together makes the combined wavefunction very stable and it is this that gives it a property called 'rigidity'. Scattering processes really do not affect the superconducting wavefunction very much because of this rigid and unbending nature and so London realized that the electronic wavefunctions in a superconductor should therefore be thought of as fixed and unresponsive objects.

As we've discussed, electrons in a superconducting wire loop can be made to go round and round for ever. Fritz London thought that such a process was rather like the motion of electrons in their orbits around an atom. This was one of the original puzzles about the behaviour of electrons in an atom which attended Rutherford and Bohr's model of the atom in 1912. If electrons orbit an atom, why do they not emit radiation (as theory says accelerating charges do) and plunge into the positively charged nucleus? Quantum mechanics provided the answer and showed that the electrons were in *stationary states* that could persist indefinitely. Might not the electrons in a superconducting wire similarly be in such a state? If this were the case, London saw that this would be quantum mechanics writ large, not on the scale of a single atom a fraction of a nanometre across, but on the scale of a piece of superconducting wire centimetres across. London therefore coined the phrase *macroscopic quantum phenomenon* to categorize superconductivity. His idea was that a macroscopic sample of superconductor behaves like a giant atom.

This led him to the realization that a superconductor could display quantum structure on a macroscopic scale. Just as the possible motions of electrons in an atom are restricted to certain quantized values, Fritz London deduced in 1948 that a consequence of his theory was that the magnetic flux (a quantity that depends on both the magnetic field and its physical extent) penetrating a superconducting loop should be quantized to certain fixed values. He calculated that the quantum of magnetic flux would be exceedingly tiny and thus impossible to observe with techniques available at his time. In fact, it was not until the 1960s, some years after London's death in 1957, that magnetic flux quantization was experimentally observed.

John Bardeen, whose story will be told in the next chapter, recalled that 1950 was a very significant year for his own understanding of superconductivity, not least because in that year Fritz London's book on superconductivity appeared. As he later wrote 'this book included very perceptive comments about the nature of the microscopic theory that have turned out to be remarkably accurate. He suggested that superconductivity requires a kind of solidification or condensation of the average momentum distribution. He also predicted the phenomenon of flux quantization, which was not observed for another dozen years.'

The London brothers, working in their house in 1930s Oxford, had made the most significant progress in the theory of superconductors in the first half of the 20th century. This however did not guarantee them a job there. Heinz London, who had the more junior position anyway, moved to work in Bristol, although eventually he returned to Oxfordshire and a job at the Harwell laboratory. Fritz had entertained hopes of staying in Oxford. However by 1936, the ICI funds which had funded the refugee scientists had dried up and Lindemann could not find funds to offer positions to all of them: he had to make a choice.

Schrödinger was a big name and was clearly a high priority to keep, though his insistence on living with both his wife and mistress raised a number of Oxford eyebrows. As it turned out, Schrödinger didn't stay and ended up in Dublin. Of the other refugee scientists, Franz Simon was a senior and respected low-temperature physicist and Kurt Mendelssohn had triumphed by successfully installing a helium liquefier in Oxford's Clarendon Laboratory, the third in the world and (more importantly to Lindemann) the first in the UK, crucially beating Cambridge. Thus both of these personnel were important for the future of the Clarendon. With a shortage of funds, no job was therefore left for Fritz London, and he was forced to move later that year, having accepted an offer of a research position at the Institut Henri Poincaré in Paris. He stayed there for three years, eventually leaving for a permanent academic position at Duke University in North Carolina. Fritz London and his wife departed from France in September 1939, though because of their German passports they weren't permitted to sail on the ship they had planned to board, being forced to take a later one. This was just as well as their intended ship was torpedoed by German U-boats.

Chapter 5
Pairing up

In the years following the Second World War, there was a large expansion in research in physics. The Manhattan project and the development of radar had convinced governments in the United States and elsewhere that what might seem at first sight to be obscure physics research could have important and quite unforeseen implications. After all, research on uranium fission had seemed a scientific curiosity in the 1930s and look how it had affected the world! Moreover, there was a growing need for highly trained scientists in industry and defence, and this fuelled an expansion in university physics departments as well as in government and industrial research laboratories. Low-temperature research benefited from this expansion, and one of the most pressing and exciting challenges around was that of the quest to understand and explain superconductivity.

The isotope effect

Theoretical physicists were making steps towards a full explanation of superconductivity, but it had so far remained elusive, despite the progress the London brothers had made along the path. One theoretician working on the problem was Herbert Fröhlich, who was trying to see if he could understand the phenomenon by including the effect of vibrations in the crystal lattice. In a crystalline solid, the atoms are located in a regular, periodic arrangement which is known as the crystal lattice. The atoms only sit perfectly still at absolute zero, but normally the atoms are jiggling around and vibrating, like hyperactive children who have been told to sit still. As you warm the crystalline solid, the jiggling and vibrations increase in amplitude. The frequency of vibration is inversely proportional to the square root of the mass of the atoms, exactly the same relationship that determines the oscillations of a mass suspended on a spring (quadrupling the mass halves the frequency of oscillation). The frequency of vibration controls the energy of a quantum of vibrational energy, and if the vibrations of atoms are in some way involved in superconductivity, you would expect the transition temperature of the superconductor to depend on the mass of the atoms.

There is a way in which this supposition can be tested experimentally. Recall that an atom consists of a very massive nucleus, containing positively charged protons and uncharged neutrons, surrounded by a cloud of negatively charged electrons. The number of protons must balance the number of electrons for the atom to be uncharged overall; however an electron is almost two thousand times lighter than either a proton or a neutron so most of the mass of an atom is located in its nucleus. The chemical identity of the atom (which element it is) is determined by the number of protons. Certain elements occur in different *isotopes*, chemically identical but having different mass. This occurs because the nucleus of the atom can accommodate a larger or smaller number of neutrons; the neutrons do not affect the charge of the nucleus but do add to its mass. For example, a normal sample of tin contains atoms with an average mass of 118.7 atomic mass units (containing a mixture of isotopes) but samples can be prepared with atoms having a mass down to 113.6 or up to 123.8, a small variation but a significant one.

With this realization, there is an obvious experiment to perform: one has to prepare isotopically pure samples of superconductors such as tin and then very accurately measure the transition temperature of each sample and see if it depends on which isotope you had used. These experiments were performed first by Emanuel Maxwell at the National Bureau of Standards and also by Bernard Serin and co-workers at Rutgers University, New Jersey, and the results were unequivocal. The transition temperatures were found to be inversely proportional to the square root of the mass of the atoms. All the pre-war theories had focused on the electrons in superconductors; the 'isotope effect' was a piece of evidence demonstrating that one should not ignore the presence of the nuclei that make up most of the mass in atoms.

The discovery in May 1950 of the isotope effect demonstrated that the vibrations of the atoms in the solid were a crucial feature in the phenomenon of superconductivity. This provided support for Fröhlich's ideas, and also for those that had been developed by another scientist working independently on the problem, John Bardeen.

John Bardeen

When theoretical physicist John Bardeen started working on superconductivity full time, he was already well on the way to his first Nobel Prize in Physics because of his invention of the first point-contact transistor. This work, performed in collaboration with experimentalist Walter Brattain, was done while both men were employed at Bell Telephone Laboratories and technically under the direction of their brilliant but prickly manager, William Shockley. It was Shockley who had first initiated research on what he had hoped would be a silicon-based field effect amplifier, but the point contact transistor worked on a completely different principle.

The publicity shots issued by Bell Labs at the time show a view of the lab with a seated Shockley, handling the new device, while Bardeen and Brattain stand behind him, cast in the role of

12. John Bardeen

subordinates (see Figure 13); in fact all the work had been done by Bardeen and Brattain with Shockley nowhere in sight. This was the beginning of Shockley's attempt to rewrite history, which he could do by his ability to control the Bell Labs public relations machine, but it was not the end of his campaign to share in the glory. He later took out the patent on the transistor based on his field effect but in his own name only; he informed Bardeen and Brattain about what he was doing by speaking to them one at a time in his office and informing them rather brusquely 'sometimes the people who do the work don't get the credit for it'. In fact, the Bell Labs patent attorneys found that there was an existing patent on a field-effect device from a physicist called Julius Lillenfeld in 1930 and so Shockley's scheming came to nothing.

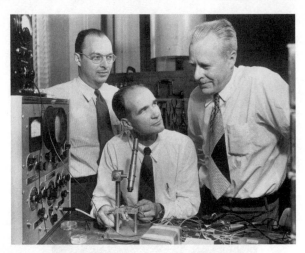

13. John Bardeen, Walter Brattain, and William Shockley (seated), in the publicity shot to highlight the invention of the transistor

Though Bardeen and Brattain had kept their boss fully informed about their progress on the point-contact transistor, the subsequent development of the junction transistor was done only by Shockley who worked alone and in secret, deliberately keeping his colleagues in the dark, and all seemingly in a fit of jealousy over Bardeen and Brattain's initial breakthrough. In fact, to ensure he had a clear run, Shockley had forbidden Bardeen and Brattain to work on the junction transistor themselves. It was the junction transistor, Shockley's invention, that turned out to lead to the dramatic revolution of electronics in the post-war period and ultimately to the development of the silicon chip and the modern computer. Bardeen, Brattain, and Shockley later shared the 1956 Nobel Prize in Physics for their invention of the transistor. Bardeen's wife, who had a hearing problem, was one of the first recipients of the new transistorized hearing aids.

However, the gentle, softly spoken, and scholarly Bardeen had had his fill of the brash, self-important control-freakery that was William Shockley and he wanted some independence. Bardeen left Bell Labs in 1951 and took up a position at the University of Illinois. Shockley himself did not last long at Bell Labs. After failing to find further promotion to the higher echelons of Bell Labs, he decided to set up his own company, Shockley Semiconductor Laboratories, in Mountain View, California, and thus founded Silicon Valley. Shockley had become convinced of his own unerring ability to spot talent (having spent some of the Second World War developing psychometric tests for hiring people) and felt that with his own proven genius (as demonstrated by his Nobel Prize) and experience of business he simply couldn't fail. However, despite recruiting the best team available to work under him (and it definitely was under him) his heavy-handed management style and unquestioning self-belief ultimately led to the company's downfall. Many of the research scientists he had employed left and formed their own companies, including Fairchild Semiconductor and eventually Intel.

Bardeen, safely at Illinois in 1951, was free to think about other physics problems and the one that absorbed him was superconductivity. He had in fact been thinking about superconductivity for a long time. Before working at Bell Labs, he had become fascinated by problems in solid state physics while doing his doctorate with Eugene Wigner at Princeton in the 1930s and, during his first academic appointment at the University of Minnesota he had an initial crack at finding a theory for superconductivity, work that was published in 1941. The war years intervened, and then the transistor work at Bell Labs, but now it was time to return to this intriguing problem.

In fact, Bardeen managed to start working on it before leaving Bell Labs; in the spring of 1950, he received a telephone call from Bernard Serin at Rutgers, informing him of the new isotope effect. Bardeen immediately dusted off his 1941 paper and updated it to take account of lattice vibrations. Herbert Fröhlich, spending the spring term at Purdue University in Indiana, was putting the finishing touches to his paper and included the late breaking news of the discovery of the isotope effect as a 'note added in proof'. Both theorists had immediately grasped the significance of the isotope effect and had constructed theories to include the effect of vibrations in the crystal lattice.

However, there was a catch. Fröhlich and Bardeen had both attempted to describe how the crystal lattice vibrations affected the energy of electrons in a superconductor, but to keep the problem even remotely tractable, the effect of the repulsion of one electron from another had been ignored. This repulsion of 'like charges' is called Coulomb repulsion, in honour of the French physicist Charles Augustin de Coulomb (1736–1806), who performed important research in electricity and magnetism. This Coulomb repulsion had been left out of Fröhlich's and Bardeen's calculations and it soon became apparent that when this was included the strong Coulomb repulsion outweighed the effect of the lattice vibrations and the models no longer worked in producing superconductivity. The Soviet theoretician Lev Landau (of whom more will be said in the following chapter) had apparently commented 'You can't repeal Coulomb's law.'

Undeterred, Bardeen attacked the problem again, this time working with a young colleague called David Pines, and together they were able to describe how, in a simplified model, electrons interact with each other in a metal if both the lattice vibrations and the Coulomb repulsion are included from the very start of the calculation. Crucially, they found that if you included the screening effect of the electrons then the Coulomb repulsion was not so much of a problem and it was possible to get an attractive interaction between electrons under certain conditions. A theoretical model was getting close, but they were not there yet.

Bardeen was also very struck by the work of a young physicist working in Cambridge by the name of Brian Pippard. Following wartime service in radar research, Pippard worked on measurements of surface resistance in superconductors at microwave frequencies and by 1950 he had begun to extract the London penetration depth as a function of magnetic field. He deduced from these measurements that there were changes in the superconductor occurring in a considerably thicker layer beneath the surface, and not just within the thin layer into which the magnetic field penetrated. Pippard proposed a new length scale, which he called the coherence length, which expresses the distance which is needed for superconductivity to turn on, measured from a region where there is no superconductivity. It is therefore not possible to have a sharp interface between superconducting and non-superconducting regions; the transition has to take place gradually over a transition layer of thickness related to the coherence length. Pippard then proposed a generalization of the London equations which were 'non-local', that is to say that they showed how the response of a particular point in the superconductor would depend not only on the magnetic field at that point (as the London brothers had assumed) but on the magnetic fields nearby, within a volume determined by the coherence length. These ideas helped to shed light on some of the new results emerging concerning the behaviour of alloys.

By the mid-1950s, Bardeen was beginning to feel that with the right type of theoretical model, it ought to be possible to deduce a superconducting state with a well-defined coherence length of the type that Pippard had proposed. But to accomplish this, it was probably going to be necessary to utilize the full machinery of quantum field theory, including the recently developed 'diagrammatic techniques' (originating in the work of Richard Feynman) which were having a substantial impact in particle physics – but this was expertise that Bardeen didn't have. David Pines left Illinois to work at Princeton and so Bardeen needed to find a new collaborator, or two. He therefore hired Leon Cooper who had been spending a postdoctoral year in Princeton. Then, with Bardeen's own PhD student, Robert Schrieffer, the three of them began to work intensively on the problem.

Cooper pairs

The initial breakthrough was Cooper's. The full problem of many interacting electrons seemed to be too complicated so, instead, he focused down on just two electrons, interacting with each other, with all the other electrons 'frozen' in place in a so-called 'Fermi sea'. In 1956, using the methods of field theory, Cooper was able to show that an arbitrarily small attraction between electrons can make it cost less energy for the two electrons to pair up together, rather than float as singletons in the Fermi sea. He therefore showed that the electron pair, now called a *Cooper pair*, is a stable entity. Cooper had shown that as long as there is some way for a weak attractive interaction to occur, even if it is extremely tiny, the system is 'unstable towards pairing', meaning that pairing of electrons will inevitably occur.

This still left unsolved the problem of what the attractive interaction might be. What causes two electrons to pair up when conventional wisdom has it that 'like charges repel' and there should therefore be a Coulomb repulsion between them? Bardeen, Cooper, and Schrieffer realized that the solution might be associated with what is called the electron–phonon interaction, that is the interaction between electrons and the vibrations in the crystal lattice. Lattice vibrations are known as 'phonons' because it turns out to be helpful to think of a lattice vibration as a kind of particle and physicists tend to give particles names ending in '-on'. As mentioned earlier, the electron–phonon interaction had already been studied by Fröhlich, and also by Bardeen and Pines. Might phonons, these vibrations of the crystal lattice, play a role in electron pairing and overcoming the Coulomb repulsion?

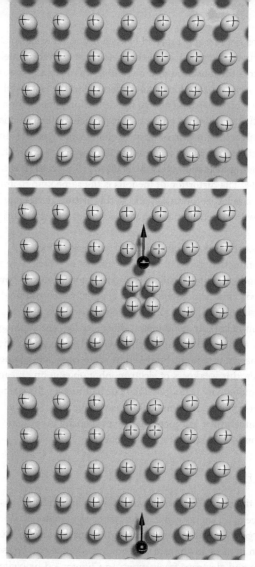

14. The lattice of positively charged ions (top panel). A negatively charged electron in a superconductor distorts the lattice of positively charged ions as it passes through (middle panel). Because the heavy ions take a longer time to respond, there is still an excess positive charge to help the second electron of the Cooper pair travel through (bottom panel)

Electrons are not only repelled by each other but are attracted to the positively charged ions in the metal, and therefore an electron will distort the ions around it by pulling them slightly towards it. The heavier ions take longer to respond than the fast, nimble electron whizzes around and so the distortion persists for a little while after the electron has left. This persisting distortion is essentially a little region of positive change and it can result in a second electron being attracted to the first electron and its surrounding distortion of positively charged ions, as shown in Figure 14. The distortion of the ions is the same kind of animal as a phonon (a lattice vibration) and so this picture gives some insight into how the electron–phonon interaction can lead to pairing.

This idea resolves one of the earliest mysteries about superconductivity: why did it occur for metals which were rather poor conductors (lead, tin, mercury) and not for metals which were good conductors (gold, silver, copper) which you might, at first sight, think were somehow 'nearer' to becoming superconductors. In good metals, the electrons and phonons interact rather weakly with each other and so the conduction of electrons is not strongly hindered by scattering from phonons. However, this weakness of electron–phonon coupling means that superconductivity is not possible. For bad metals, the strong electron–phonon coupling makes scattering more likely at high temperature, but increases the propensity to form a superconducting state at low temperature.

Mind the gap

One of the big differences between metals and insulators is the presence of an energy gap in the latter class of material. What is an energy gap? This is the energy cost you must pay if you want an electron to move in an insulator. For a metal, the cost is essentially nothing and electrons freely move. In Chapter 1, I likened the movement of a gas of electrons in a metal due to an applied voltage to the wafting of a cloud of bees by a gentle breeze. Even if the breeze is very, very gentle, the cloud of bees will still drift along, even if very slowly. In contrast, the electrons in an insulator are much more like limpets stuck fast to a rock. These limpets are impervious to a gentle breeze, or even to a strong wind. If you want a limpet to move, you have to pay an energy cost to forcibly eject it from the rock (this is only a 'thought experiment'; no limpets were harmed during the making of this analogy). This effort in unfastening the limpet is analogous to the energy gap in an insulator.

Since superconductors are much closer to metals than to insulators, you might expect that they would not possess an energy gap. But you would be wrong. Rolfe Glover and Michael Tinkham exposed superconductors to infrared radiation in an elegant experiment in 1956 and showed that if the energy of the infrared radiation was above a certain threshold value, the superconductors absorbed radiation very effectively, but if it was below the threshold value then they did not. This is very good evidence that the carriers in a superconductor behave as if they do have an energy gap. What is going on?

The idea of a Cooper pair offers a way of understanding this. The superconducting gap energy is the penalty you have to pay to break up a Cooper pair. Since the pairs are bound, it takes a certain amount of energy, called the binding energy, to break the pairs up and this leads to what is known as the superconducting gap. In a normal metal, you can give electrons an arbitrarily small amount of energy to excite them; in a superconductor, nothing will happen until you supply an amount of energy equal to the binding energy and once you have bridged that gap then that energy can be absorbed. This effect can be measured by looking at the way in which superconductors reflect electromagnetic waves; if the waves have an energy (determined by their frequency) which is smaller than the gap energy, the waves are not absorbed and reflect straight back from the superconductor; however, as soon as the energy is large enough, superconducting pairs can be broken apart and energy is absorbed.

The BCS theory

Though Cooper had cracked the mechanism of the pairing of electrons, there was still some way to go before a full understanding of the superconducting state could be achieved. Cooper had only worked with a single pair of electrons in an infinite metal. It was not obvious how you could generalize this to all the electrons in the metal.

A further hitch arose. While Cooper's result was highly suggestive, the BCS team worked out the number of expected pairs of electrons in a superconductor. A pair of electrons is described by a wavefunction, and this wavefunction was supposed to be coherent over a small volume with a size given by Pippard's coherence length, which is essentially the average distance between the electrons in the pair. However, within that small volume they estimated that around a million other pairs would be found. This was bad news for making a simple theory. They would prefer that the pairs were well separated so that they could construct a model of the equivalent of a rather empty dance floor with the occasional couple foxtrotting in locations here and there. That situation was so much easier to treat theoretically. What they actually had was a jam-packed dancefloor heaving with a throbbing multitude of dancers, but the two members of an individual dancing pair might each be on opposite sides of the room! What on earth would keep the individual dancers in these highly separated pairs dancing in step with each other when there were so many other dancers in the way? Returning to the electrons, their worry was that the pairs would overlap so strongly that the mechanism of getting them all to join together into a superconducting state would be interrupted by all the collisions between pairs and this would break the pairs apart.

The crucial breakthrough came to Robert Schrieffer, and apparently it hit him while he was thinking about the problem on a New York subway train. Suddenly he realized how to write down the wavefunction which describes the superconducting state and astonishingly it required considering all the electrons in the material together. No longer could one rely on the independent-electron model which was the mainstay of existing solid state physics whereby you could consider electrons one at a time, as if each behaved as an independent entity. In the superconducting state, an enormous number of electrons acted in concert, as if each was part of a larger, inseparable whole. Schrieffer had discovered what has become known as the BCS wavefunction, named in honour of the three scientists, Bardeen, Cooper, and Schrieffer, who had cracked the problem.

In constructing this wavefunction, Schrieffer had used a number of tricks. One of these was to write down an expression which, unusually, did not contain a fixed number of electrons. How can

15. 'BCS' (alias John Bardeen, Leon Cooper, and Robert Schrieffer, obligingly standing in their correct order)

this be? Recall Schrödinger's cat, which is famously both alive and dead. This illustrates one of the strange features of quantum mechanics that systems can exist as a superposition of different states with different properties, so that things can exist quite literally as a sum of contradictions. Schrödinger's cat of course points out the absurdity of a description that works for atoms but seems insane when you are discussing something large and macroscopic, like a cat. Nevertheless, this quantum superposition of contradictions seems to describe the nature of superconductors very well. Schrieffer's wavefunction was just such a strange quantum superposition of ghostly incarnations of the superconductor with different numbers of electrons. Every time you try and ascertain how many electrons there are in this wavefunction, you get a different answer, as electrons bob in and out of existence. This at first sight seems very odd: has a piece of real superconductor not got a fixed number of electrons? It has, but Schrieffer found it expedient to treat a superconductor as if it were connected to an electron reservoir so that the number of electrons in the superconductor could vary. He realized that a more crucial requirement for the superconducting state was that it should have a fixed *phase*.

It was the fact that the phase of the wavefunction describing all the superconducting electrons locks into a single value which was the defining quality of a superconductor and Schrieffer wanted to get that written in right at the start. The BCS wavefunction was therefore what is known as a *coherent state*, a type of quantum mechanical object which had been invented by Schrödinger 30 years previously (he called it a minimum uncertainty wave packet), and it is also used in describing the properties of laser radiation.

Schrieffer knew the kind of wavefunction he wanted but did not know precisely how to write it down. He therefore used a trick which is very standard in quantum mechanics: he included a number of parameters in the model whose value he did not know, and then proceeded to find their value by working out the energy of this state and minimizing the energy with respect to those parameters. In a sense, his theory had a sufficient number of degrees of freedom that these could be adjusted to get the best fit to reality and, on the principle that Nature finds the lowest-energy solution to a problem, his wavefunction could be adjusted to provide an optimal solution. So the BCS wavefunction was a guess, but an inspired guess and one which was so constructed that it could be easily adjusted to yield an even better one.

Deducing the wavefunction was not enough. What was needed was to show how this wavefunction explained all the properties of the superconducting state. Bardeen, Cooper, and Schrieffer divided up the various aspects of the problem between them: the thermodynamic consequences, the effect on nuclear magnetic resonance experiments, the implications for transport properties, the Meissner effect and so on. All of these outworkings of the theoretical framework had to be gone through, and much to their delight, Bardeen, Cooper, and Schrieffer found that they were able to systematically produce predictions of the experimental properties that matched completely with what had been found in real experiments. The BCS theory worked. Their paper was published in *Physical Review* in 1957 and was recognized immediately as a masterpiece. The BCS theory explained most of the properties that had been observed in superconductors up until that time and in 1973 would win its inventors a Nobel Prize. Bardeen became the first person to win two prizes in the same field, and at the time of writing is the only person to have achieved this feat in physics. The BCS theory is rightly regarded as the one of the major triumphs of 20th-century theoretical physics.

Many body

How does the BCS theory explain superconductivity? The first thing to realize is that in superconductors, it is not that the electrons suddenly become immune from scattering off impurities and off each other (the effect that gives electrical resistance in a normal metal), it is simply that scattering does not affect conduction. The two electrons in a superconducting pair have equal and opposite momentum with respect to each other, although the pair as a whole can glide through a sample when a current flows. All the pairs are locked together and glide at the same speed. If a pair scatters with a phonon, then the two electrons may change their individual momenta but the gliding of the pair along the current direction continues completely unaffected. In effect, a pair state scatters into another pair state. This occurs because there is a very great energy saving for keeping all pairs moving with identical speeds. It costs too much energy to knock a pair out of line with all the others. The electrical current represented by this uniformly gliding assembly of electrons, once started, can continue forever.

Moreover, it is not really correct to think of a single pair in isolation. The BCS state is what is known as a 'many-body' state in which it is illegitimate to think of the system as simply a number of individual states; there is no way to avoid thinking of the whole lot together. Indeed, the very strength and nature of this state is derived from the way in which the pairs interact together. The scattering events which change one pair into another are the very interactions which fortify the BCS state. One is perhaps reminded of the words of the Gerasene demon: 'We are legion, for we are many'. A more modern representation of the superconducting state is provided by the *Terminator* films that star Arnold Schwarzenegger; the evil robot is hell-bent on striding towards his quest of the young John Connor and even when shot, flattened, blown up or incinerated, his molten remains begin to coalesce and reform so that the Terminator emerges from the flames apparently unscathed and with the same striding momentum as before. The BCS state is like that. Honestly.

This has consequences for the superconducting energy gap. The size of the gap turns out to measure the strength of the superconducting state and this depends on the number of pairs in the BCS state. As you warm up a superconductor, it becomes possible for pairs to be broken up into single electrons: the thermal energy must supply the energy needed to break a pair. As you warm it up, there is more thermal energy around and it becomes easier to find the energy to break a pair of electrons. However, an additional effect occurs: as more and more pairs are broken up, the superconducting state weakens and this reduces the size of the energy gap. This makes the energy price to break a pair smaller and more pairs are broken. Eventually, at a sufficiently high temperature, the size of the energy gap shrinks to zero and no energy is now needed to break a pair and there is no incentive for electrons to pair up. Superconductivity is destroyed and normal metallic behaviour resumes. The temperature at which this happens is the superconducting transition temperature.

The BCS theory is not based on what the attractive interaction between electrons actually is, it just assumes that there is one. It is therefore more general than is first supposed and could apply to more general cases than just electrons pairing up because of lattice vibrations.

The BCS theory was the product of three incisive minds, but one has to be particularly in awe of Bardeen, winner of two Nobel Prizes and co-inventor of the transistor as well as the BCS theory. His name is virtually unknown in popular culture and yet his impact has been extraordinary. Of course, he didn't have the crazy hair of Albert Einstein's later years or the wild and effervescent personality of Richard Feynman; the softly spoken Bardeen looked a bit like a bank manager, played golf, and lived rather quietly. It is a shame that gentle modesty gets a poor press; on the day he and Brattain invented the transistor, he returned home in the evening and mumbled to his wife 'we discovered something today'. He didn't say what, or go into further details, but by those few uncharacteristically effusive words she knew that whatever he had discovered, it must have been important. I rather like that.

Chapter 6
Symmetry

The BCS theory of superconductivity looked like the final word on the subject. It showed how electrons could pair up and how these pairs could together form a giant collective unit in which the phases of all the individual wavefunctions locked together. This theory explained all the phenomena observed so far. However, as so often occurs in physics, there was a completely different approach to the problem that would turn out to be just as illuminating. It was only after both approaches had been developed that it was shown that they were completely consistent with one another. BCS had come from the United States, but this second approach was developed in what was then the Soviet Union. It starts with a deliberate decision not to worry about the microscopic details of the problem (such as those with which BCS were concerned), the trees if you like, but to look at the big picture: the wood. This was just the sort of viewpoint loved by the remarkable physicist Landau.

Dau

Lev Davidovich Landau (or 'Dau' as he was known to his students and colleagues) was born in 1908 in Azerbaijan and was something of a child prodigy. After a spell working outside the Soviet Union, particularly in Copenhagen with Niels Bohr, the father of quantum mechanics, Landau headed the department of theoretical physics at Kharkov from 1932 until 1937 when he moved to Moscow as head of the Theoretical Division at the Institute for Physical Problems. Following the move to Moscow, Landau was arrested in April 1938 as part of Stalin's repressive purges and spent a year in prison. At that time, more than a million people were arrested and hundreds of thousands were executed. Landau was lucky to be released, though this was said to be as a result of the intervention of the low-temperature physicist Pyotr Kapitsa, who wrote a pleading letter to Stalin on Landau's behalf.

Landau set up a famous school of theoretical physics which was used to train all the best Soviet physicists, an exceptionally tough apprenticeship culminating in an exam which Landau called the 'Theoretical minimum'. The bar was set exceptionally high, and only 43 students ever attained the theoretical minimum; those that did went on to positions of high eminence. Landau also began work on a ten-volume *Course of Theoretical Physics* with his colleague E. M. Lifshitz. This vast work spans the entire range of physics, with each result and equation presented and derived in a highly original way, bearing all the hallmarks of Landau's instinctive and piercing insight. The books are still highly valuable (particularly the early volumes in which Landau had more direct input), though they are extremely heavy-going and generally of value only when covering a topic you already know!

Landau was a merciless interrogator of lesser mortals, which included pretty much everyone, and to give a seminar at Landau's institute and in his presence was said to be a terrifying experience. Outside his office in the Ukrainian Physicotechnical Institute was a nameplate which bore the inscription:

L. LANDAU

BEWARE, HE BITES

Nevertheless, Landau was held in awe and affection and he had a defining, personal influence on the development of Soviet theoretical physics.

16. Lev Landau

Landau enjoyed classifying 20th-century theoretical physicists according to a logarithmic scale of his own invention. This scale was logarithmic to the base ten, so that a first class physicist was ten times better than a second-class physicist, who was in turn ten times better than a third class physicist. On this scale, Einstein scored 0.5, Bohr, de Broglie, Dirac, Feynman, and Heisenberg got class 1. Pauli slipped to 1.5. Landau put himself in class 2.5, only later upgrading himself to class 2. Of course most physicists whom Landau came across would count themselves lucky to get into class 3 or 4. Landau was immensely prolific and worked in many fields of theoretical physics, but perhaps his best-known work amongst physicists is his contribution to the theory of phase transitions.

Phase transitions

When an ice cube melts, or the water in a kettle boils, there is said to be a phase transition. Solid ice changes to liquid water, or liquid water changes to gaseous steam. Phase transitions are rather unusual phenomena: water at $30°C$ is not very different to water at $28°C$, it's just a bit warmer. The molecules wiggle around a bit faster, but there is a continuum that exists between water at $28°C$ and water at $30°C$. However, there is something very, very different between water at $1°C$ and ice at $-1°C$. The two things are entirely different. Passing through the phase transition at $0°C$, when water freezes, changes its nature discontinuously.

Another example is what is known as a ferromagnet, which is a material which can exhibit spontaneous magnetism. A good example of a ferromagnet is iron. Iron can display spontaneous magnetism: all the tiny atomic magnets in a piece of iron can point in the same direction (the left-hand picture in Figure 17), making the piece of iron magnetic; paper clips will stick to it! However, if you warm the piece of iron up to very high temperature (above 1043K, which is $770°C$), it will lose its magnetism when you pass through the ferromagnetic transition temperature (called the Curie point, in honour of the French physicist Pierre Curie, whose more famous wife, Marie Curie, is known for her discovery of radium). At high temperature, all the little atomic magnets will point randomly (the right-hand picture in Figure 17). This change between magnetic and non-magnetic states at the Curie point is another example of a phase transition.

In the 1930s, Landau made a very important contribution to the understanding of phase transitions. In essence, Landau's great insight was to see that you didn't need to worry about the microscopic details of what was going on in a phase transition. Landau decided instead to ask what kind of theory might describe the phase transition and decided that the right way to think about

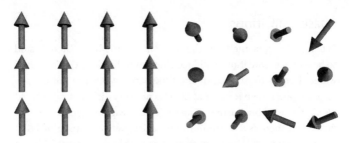

17. A ferromagnet at (left) low temperature and (right) high temperature

it was in terms of symmetry. He knew that if a system is at equilibrium, then its energy takes the lowest possible value (a ball rolls down to the bottom of the hill all on its own, but never up to the top), so if he could find a formula for the energy, all he had to do was find its minimum value. He made a guess for the general formula for the energy of a magnet by trying to find the simplest formula that would do the job. When you try this kind of approach, it turns out that the symmetry of the problem restricts your choice of the type of formula you can use.

Landau's result is shown in Figure 18. At high temperatures, the stable state of the system (identified by the point of minimum energy) is exactly where a ball would roll to if released on a surface shaped like the curve in the upper figure. This puts the system in the symmetrical position in the middle, which in Landau's model corresponds to the value of some physical property, such as magnetism, being zero. No direction is singled out as being special, and this solution has the full rotational symmetry. It causes the little atomic magnets (the arrows on the right-hand side of Figure 17) to point in any direction they like.

At low temperature, the energy is now shaped like the right-hand curve; the stable point of the system is now located either displaced to the right or to the left. This means that the system can either be

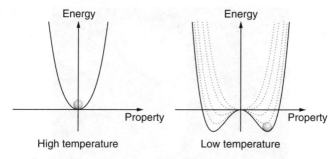

| High temperature | Low temperature |

18. The Landau model of phase transitions

magnetized in one direction or the other, but at any rate it will be
magnetized. However, its freedom to choose one direction of
magnetization rather than any another means that, to use the
technical jargon, it 'breaks the symmetry', which in this case is
rotational symmetry. Now we have chosen one particular direction
to be 'special' and the symmetry of any direction being the same has
been broken. Landau therefore had shown that symmetry breaking
occurs at this phase transition.

Ginzburg and Landau

One of Landau's collaborators was Vitaly Lazarevich Ginzburg,
eight years Landau's junior. Ginzburg had been working on
superconductivity in the mid-1940s, particularly in generalizing
London's work. However, in the late 1940s he became preoccupied
with other problems, rather productively in fact, so that he made
major advances in radio-wave propagation in the ionosphere, the
theory of ferroelectrics, light scattering in liquids and a host of
other topics. Ginzburg was very impressed by Landau's work on
phase transitions and had been thinking about how to apply it to
the phase transitions inside superconductors. The two of them,
Ginzburg and Landau, worked on the problem together and the
result, now known as the Ginzburg–Landau theory, was published
in 1950, seven years ahead of the BCS theory.

19. Vitaly Ginzburg

Their approach was to apply Landau's model to superconductors and to consider the energy in terms of a property to which they gave the symbol $|\psi|^2$ which represents the number of superconducting carriers per unit volume. The equations showed that there were no superconducting carriers above the critical temperature, but they appeared below (see also Figure 20).

Ginzburg and Landau also included some extra terms in their expressions for the energy. One term included the fact that it saves energy if the superconducting carriers spread out uniformly and it costs energy if there are sudden changes in the number of carriers from one place to another; once again, they guessed the simplest formula which made this work. Next, they remembered another important symmetry principle which goes by the very grand-sounding name of *gauge symmetry*. This is a very general concept in physics, but it is possible to give a very simple example of it. In an electrical circuit, one point is often labelled as 'ground', and this corresponds to the third pin on a three-pin electrical mains plug.

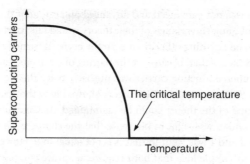

20. Superconducting order appears below the critical temperature

In the circuit, ground is at zero volts. However, since in circuits you only ever measure 'potential differences', it would be quite possible (though perverse) to redefine 'ground' to be 40,000 Volts and have all the other voltages in the circuit defined with respect to that. This looseness in the description of electricity (the fact that you can define the electrical ground to be whatever you like) is an example of gauge symmetry. To preserve this symmetry, Ginzburg and Landau had to write down their equations in a particular way. When they did this, they ended up with a pair of rather complicated differential equations for the quantity ψ, but very encouragingly they found that the solutions of these equations obligingly spit out the London equations, the penetration depth and the coherence length, in other words a whole lot of physics that had taken a couple of decades for others to formulate.

Because their theory was phenomenological, the charge of the superconducting carrier, which they called e^*, could be taken to be anything. Landau didn't see why it should be any different from the charge of the electron e, so in the paper they explicitly wrote 'there are no grounds to believe that the charge e^* is different from the electron charge'. Five years later, Ginzburg realized that agreement with experimental data could be achieved if they took e^* as somewhere around twice or three times the electronic charge.

Landau was not convinced and advanced an argument that the beautiful 'gauge invariance' of their theory would fall apart if e^* was taken to be anything other than e. In the event, it turned out from the BCS theory that, because of the pairing of electrons, the charge of the superconducting carrier was precisely twice the electronic charge. Furthermore, the BCS theory showed how the 'gauge invariance' of the theory could be maintained. As Ginzburg put it later, 'Landau was right in the sense that the charge e^* should be universal and I was right in that it is not equal to e. However, the seemingly simple idea that both requirements are compatible and $e^* = 2e$ occurred to none of us.' He also lamented that he did not see the solution that Bardeen, Cooper, and Schrieffer had so clearly grasped.

Nevertheless, Lev Gor'kov showed in 1959 that the Ginzburg–Landau equations could be derived from the BCS theory, and the Ginzburg–Landau approach is much less unwieldy for deriving important aspects of superconductivity. For Landau's many achievements, he was awarded the Nobel Prize in 1962. Unfortunately, he was not able to collect it. Earlier that year, a car accident on an icy road between Moscow and Dubna left him in a coma for several months and he never properly recovered, dying six years later. Ginzburg had to wait a very long time for his Nobel recognition; aged 87, he collected the prize in 2003.

Alloys and the 'dirt effect'

Physicists often like to start with the simplest systems and therefore, when faced with a new phenomenon like superconductivity, begin to focus in on the chemical elements. Once you start mixing up different elements, things get complicated, so why bother? Moreover, it was known that in ordinary metals if you have impurities in a sample, this leads to extra resistance (known as 'residual resistance') and this departure from pure behaviour looks like nothing but a nuisance. In this respect, Wolfgang Pauli typified the approach of theoretical physicists. Writing to his assistant

Peierls, he declared 'the residual resistance is a dirt effect and in the dirt one should not stir'.

However, it is well known that mixing up elements can give you something that is more than the sum of its parts. A trivial example is water (H_2O) which is much more than a simple mixture of hydrogen and oxygen. Early in human history, the discovery of alloys led to a technological revolution. An alloy is a homogeneous mixture of elements, at least one of which is a metal. For example, brass is an alloy of copper and zinc, bronze is an alloy of copper and tin (though sometimes with other elements in as well). Steel contains iron, carbon, and various other ingredients. This is definitely 'stirring in the dirt', but without these alloys civilization could not have developed. It was a natural step to study superconductivity in alloys, and when people did they had a big surprise.

Various groups rose to the challenge of studying superconducting alloys, including those of Mendelssohn at Oxford, Shoenberg in Cambridge, de Haas in Leiden, and Shubnikov in Kharkov. Alloys appeared to behave very differently from elements and, in particular, certain alloys had much larger critical magnetic fields (the magnetic field to destroy superconductivity) than found in elements, though there was a puzzle that they did not seem to completely exclude magnetic fields (the Meissner effect) in the way that the earlier superconductors did. The large critical field was a potential breakthrough because if a superconductor could withstand a larger magnetic field, then it can carry more electrical current before superconductivity is lost. This revived (correctly as it turned out) the old hope entertained by Onnes that one might be able to make superconducting magnets, though the future seemed to be with wires made from a suitable superconducting alloy.

The earliest measurements of the magnetization of superconducting alloys had been done in Leiden in 1935, but there were worries about whether the samples they had made were of sufficient quality and properly homogeneous (well mixed). Lev Shubnikov had worked with de Haas in Leiden in the 1920s and had established a low-temperature laboratory in Kharkov in 1930; there he made better samples, heating his alloys a long time at temperatures close to the melting point to make them as homogeneous as possible (a process called annealing). He studied the magnetic properties of his samples in detail and showed that they responded to magnetic fields in a manner completely different to elements. Because his samples were so clean, he was convinced that this was a real effect and not an artefact. Unfortunately, Shubnikov did not survive to see the fruits of his work; in the same purges that had caused Landau's arrest, Shubnikov was falsely accused of attempting to organize an anti-Soviet strike, arrested and executed in 1937. He was 36 years old.

More than a decade later, in the early 1950s, Alexei Alexeyevich Abrikosov was working at the Institute for Physical Problems of the USSR Academy of Sciences and had been very impressed by the Ginzburg–Landau theory. Nevertheless, he was concerned that some data measured by one of his experimentalist friends did not seem to fit the theory. He therefore realized that the theory needed to be extended into a regime in which Ginzburg and Landau had not imagined it could be taken.

To understand Abrikosov's argument (and if you don't want to, now is a good time to skip to the next section) one needs to understand the balance of energy in a superconductor. At low temperature, the electrons prefer to condense into pairs and make the superconducting state, and this is because it costs them less energy to do so. There is a quantity of energy they save, which we will call the superconducting condensation energy. Recall that superconductivity can be destroyed by a magnetic field, but that the magnetic field penetrates a certain distance into the surface. Since excluding magnetic field costs energy, this small penetration of the magnetic field represents a bit of energy saving. However, in the bulk of the superconductor this is more

21. Alexei Abrikosov

than paid for by the superconducting condensation energy saving.

Furthermore, the superconducting wavefunction cannot change abruptly because this costs energy. Hence, the superconducting wavefunction must decay to zero as you approach the surface over a length similar to the coherence length. This leads to an energy cost because over this distance the system fails to save its superconducting condensation energy. In the first superconductors to be discovered (mercury, lead, tin, etc.), the coherence length is much larger than the penetration depth and so the energy cost of destroying superconductivity near the interface outweighs the energy bonus of allowing the field to penetrate a bit. What this means is that the interface between the superconducting state and the normal state is costly and the system will prefer not to make an interface unless it has to.

Abrikosov called these traditional superconductors (mercury, lead, tin, etc.) type I superconductors, to distinguish them from type II superconductors to be discussed now. In a type II superconductor the situation we have just described is reversed. The penetration depth is now much longer than the coherence length, and so the energy cost of destroying superconductivity near the interface is dwarfed by the energy bonus of allowing the field to penetrate. This means that having an interface between superconducting and normal states saves energy, and so the formation of interfaces is extremely favourable. A type II superconductor is going to be full of interfaces!

This means that in a type II superconductor the normal state and superconducting state become as finely divided as possible. Abrikosov was able to show that the magnetic field penetration will now occur in single tubes of non-superconducting (normal) material. Each tube contains a quantum of magnetic flux and electrical current flows around each tube to shield the superconducting region around them from magnetic field. These tubes are called vortices because of the way the electrical current circulates around them (see Figure 22). Abrikosov was able to show that the vortices provided the explanation for the observation that many alloy superconductors appeared to exhibit an imperfect Meissner effect by allowing magnetic flux to penetrate through them. His model also provided excellent agreement with Shubnikov's experimental work on the magnetization of alloys back in the 1930s.

The vortices repel each other and arrange themselves into a regular arrangement. Abrikosov had guessed that the vortices would arrange themselves into a square two-dimensional lattice. But in

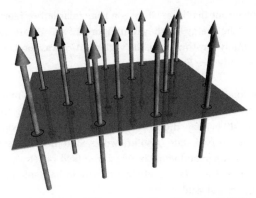

22. Schematic diagram of the vortex lattice. Magnetic field lines (the arrows) are arranged in a triangular formation. Currents circulate around the vortex lines to screen the rest of the superconductor

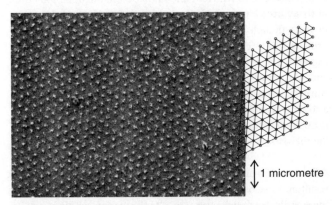

1 micrometre

23. The vortex lattice in the superconductor MgB$_2$ observed using a magnetic decoration technique. A sketch of the triangular lattice is shown to the right

fact, it turns out that in most cases a triangular lattice more often minimizes the energy (as shown in Figure 22 and Figure 23). Encouragingly for Abrikosov, the vortex lattice was soon observed experimentally (see Figure 23).

We now know that type II superconductivity is much more common than type I superconductivity; the latter being the exception rather than the rule. Abrikosov now understood that the so-called mixed state (also known as Shubnikov phase) of a type II superconductor, in which the field penetrates the superconductor as a lattice of vortices, would be stable up to a large critical field.

Abrikosov worked out his ideas about a lattice of superconducting vortices in 1953, but Landau was not convinced and thought that wild imaginings of vortices smacked of 'pseudoscience' and so Abrikosov held back in publishing. However, two years later the American physicist Richard Feynman explained some of the properties of very low-temperature liquid helium (in what is known as its superfluid state) and described the vortices existing in it. This work convinced Landau that there might be something in this vortex idea and Abrikosov's work finally saw publication in 1957, though initially only in Russian. Even after being translated into English, it attracted little attention until more experimental work on alloys was done in the West in the 1960s and Abrikosov's vortex lattice could be observed experimentally. Abrikosov shared the 2003 Nobel Prize with Ginzburg and also with Anthony Leggett, a physicist who had made major contributions to the theory of superfluidity (but that is another story).

When a supercurrent flows in a type II superconductor, there is a resultant force on the vortices which acts in a transverse direction (perpendicular to the current and to the vortices). This causes dissipation due to the normal material in the cores of the vortices and results in electrical resistance, exactly what superconductors are supposed to avoid! This is bad news for practical applications, but fortunately there is a solution. If the superconductor contains suitable impurities, these can pin the vortices in place and stop them moving as an electrical current drifts past them. This pinning effect turns out to be vital for making type II superconductors useful, and once again it is the presence of Pauli's hated 'dirt' effect that has come to the rescue.

The Higgs boson

The Ginzburg–Landau approach showed that superconductivity involves a strange and profound effect which goes by the name of '*spontaneous symmetry breaking*'. This helps to explain some aspects of the Meissner effect, which you will recall is the expulsion of magnetic fields by a superconductor. This arises because of the way in which the phase of the macroscopic wavefunction locks onto a single value (this 'breaks' a symmetry, because previously the phases of individual wavefunctions were free to take any value) resulting in the electromagnetic forces, whose influence is usually very long-ranged (which is how your television and mobile phone work), becoming short-ranged inside the superconductor. In fact, the equation describing the magnetic field inside a superconductor looks like the electromagnetic wave equation written in such a way so as to include photons having mass. Now photons do *not* have mass, which is why they travel at the speed of light, but inside a superconductor the close coupling of current and magnetic field (discovered by the London brothers) means that photons behave *as if they do have mass*. This gives rise to the short-range electromagnetic forces, the appearance of currents on the surface of a superconductor which screen the interior from magnetic field, and hence the Meissner effect.

The sudden appearance (and it is only an appearance) of mass comes from the symmetry breaking that is inherent in the low-temperature diagram in Figure 18. The new minimum that the system sits in (the ball, displaced to the right in that diagram) corresponds to the spontaneously symmetry broken state that doesn't possess the symmetry that you have at high temperature (the ball sitting in the original central minimum). The Ginzburg–Landau approach shows in detail that it is interaction between electromagnetic fields and the superconducting carriers that determines how the superconductor responds to any disturbance, and produces screening of magnetic fields.

Philip Anderson, then at Bell Laboratories, wondered if the physics behind superconductivity had more general applicability, and this led to the prediction of what is now called the 'Higgs boson' (often nicknamed the 'God particle') which is currently being looked for at the Large Hadron Collider (LHC) in CERN. Peter Higgs, at Edinburgh University, made the decisive step in the particle's prediction, but would be the first to admit that several others also made crucial contributions. It probably should be called the Anderson–Nambu–Higgs–Brout–Englert–Guralnik–Hagen–Kibble boson, but for some reason that name just hasn't caught on.

The idea (very roughly) is that all the mass in the Universe appears in much the same way that a photon appears massive in a superconductor. The whole Universe is supposed to be permeated by Higgs bosons, in much the same way as a superconductor is filled with superconducting pairs. This all-pervading bath of Higgs bosons is called the Higgs field. Particles that do not interact with the Higgs field are able to travel through the Universe unimpeded (like photons, which travel through empty space at the speed of light). Many other particles (such as electrons and quarks) do interact with the Higgs field, and as a result acquire mass. Thus the Higgs boson is proposed as the particle that explains the occurrence of mass in the Universe. In a sense, the Universe in which we live behaves like a giant superconductor!

At the time of writing, the LHC has started taking data in its search for the elusive Higgs boson. Though they are purported to permeate the Universe, it is only possible to observe Higgs bosons directly in high energy collisions. It is therefore somewhat fitting then that the experiment to search for the Higgs boson is using huge numbers of superconducting magnets to steer the beam around the ring.

Chapter 7
Before the breakthrough

During the 1960s and 1970s, superconductivity research went through a period of consolidation. The theoretical landmarks of the BCS and Ginzburg–Landau theories had been passed, and it was a time to work out their consequences. It was also a lull before the extraordinary breakthroughs of the late 1980s, which we will describe in the next chapter, when many preconceptions about superconductivity were to be upset. But it was nevertheless a time when superconductivity came of age. Two crucial advances were made that led to superconductivity at last being useful. These advances were the discovery of the Josephson effect and the development of techniques to synthesize new materials. They will be described in turn in this chapter.

Tunnelling

Despite the famous Monty Python sketch about the Society for Putting Things on Top of Other Things ('This year, our members have put more things on top of other things than any year before'), there is something to be gained by putting things on top of other things. Sandwich structures or multilayers form much current technology providing the lasers in our CD players, the processors in our computers, and the sensors in hard disks. It wasn't long before people tried to make layers of superconductors and sandwich them between other materials.

In 1960, Ivar Giaever at General Electric Laboratories in the US made the first superconducting tunnel junction. Giaever had been trained as a mechanical engineer and later claimed that he had only been hired by General Electric because of their lack of understanding of the Norwegian grading system! Giaever's superconducting tunnel junction consisted of two superconductors separated by a very thin insulating layer. By making an electrical circuit with this tunnel junction it was possible to see how electrons could flow through the insulating layer, an effect which is impossible in classical physics. This is accomplished because of the ability of electrons to perform a ghost-like process of *tunnelling* through the insulating barrier, much as a phantom can supposedly pass unhindered through a solid wall. The apparently mysterious tunnelling process is in fact well described by quantum mechanics and the electrical characteristics of the superconducting tunnel junction can be used to understand the properties of the superconducting layers (they were used to infer the existence of an energy gap described in Chapter 5).

The weakest link

Brian Josephson was a doctoral student at Cambridge University's Cavendish Laboratory in the early 1960s, working under the supervision of Brian Pippard (the originator of the coherence length). During the first year of his doctorate, he had taken some lectures from Philip Anderson who was at that time spending part of every year in Cambridge. Anderson lectured on broken symmetry as a central principle underlying solid state physics and Josephson was captivated by these ideas. He began to appreciate that the breaking of symmetry in a superconductor was really its fundamental defining quality and started to think what observable consequences there might be. This was a difficult question to answer because the superconducting state had a unique phase, but

24. Brian Josephson

the standard wisdom in quantum mechanics was that the phase of a wavefunction was an arbitrary quantity which one couldn't measure.

Josephson had identified a very important problem and this led to a brilliant insight. He realized that though the phase of the wavefunction inside a superconductor was fixed and uniform inside it, the phase of the wavefunction inside a second superconductor would also be fixed and uniform, but would be fixed at a different value from the first. If these two superconductors were brought in close proximity to one another, perhaps separated by a very thin non-superconducting barrier (known in the trade as a 'weak link'), then the phase difference between them would have observable consequences. He performed a calculation of the quantum-mechanical tunnelling current between the two superconductors and found that a spontaneous net current would flow from one to the other which was directly related to the difference in the values of phase taken by the two superconductors.

This was a totally unexpected result: first, that the quantum mechanical phase of a wavefunction should have an observable effect and second, that a spontaneous current would flow between two superconductors. This remarkable prediction, made in the first year of his doctorate, was later to win Josephson a Nobel Prize, but Josephson's PhD supervisor was not convinced it was of sufficient worth to win him his doctorate. Josephson therefore spent the second year of his doctorate trying to provide an experimental confirmation of his prediction, a task that neither suited his own skills nor the facilities of his laboratory. It was not a trivial matter to construct what is now known as a Josephson junction, two superconductors connected through a weak link, and far more difficult than a conventional tunnel junction such as was made by Giaever, which has a more insulating barrier. Philip Anderson, who had been closely involved with the development of Josephson's thinking, and who had agreed to give Josephson a year to produce experimental justification before competing with him, eventually constructed a working Josephson junction himself at Bell Labs in collaboration with John Rowell in 1963.

A spontaneous current is all very well, but the Josephson junction has more tricks up its sleeve. Josephson had also realized that if a steady voltage were to be applied across such a junction, then it would spontaneously oscillate and an alternating current would be produced. This occurs because it causes the quantum mechanical phases of the superconducting wavefunctions on each side to precess at different rates, thus driving an alternating Josephson current. Another way of understanding this process is as follows: as we have seen, superconducting currents can flow across a Josephson junction and because these are superconducting currents, there is no dissipation of energy; hence, by applying a

25. Philip Anderson

voltage one is giving the electrons energy that they do not need and cannot easily dissipate; the result is that the electrons oscillate back and forth across the junction, radiating energy as electromagnetic waves. This is known as the *a.c. Josephson effect* (a.c. stands for alternating current) and it was soon experimentally demonstrated and had far reaching consequences.

The first of these is that the typical frequency at which the oscillations occur is in the microwave region (a voltage of 1 mV produces a frequency of 486 GHz); thus a Josephson junction can be used as a generator of microwaves. It turns out that a single junction produces only a very small microwave power, but it was soon possible to produce arrays of Josephson junctions which produce more power.

A second use comes about because frequency is a quantity which is very easy to measure accurately and to control. Therefore, the Josephson effect can be used to give a new standard of voltage. For a long time, the best standard voltage was known to 1 part in 10^6 and was given by a particular type of electrochemical cell that needed to be stored in a temperature-controlled room and from which all voltmeters were ultimately calibrated. Today, such cells have been replaced by arrays of Josephson junctions which provide a standard volt which is accurate to one part in 10^{12}.

Furthermore, the link between frequency and voltage provided by the Josephson effect involves two fundamental constants, the charge on an electron and Planck's constant. The ratio of these two constants is today best known via measurements of the very same array of Josephson junctions. Thus Josephson's musing about symmetry breaking led to a crucial breakthrough in metrology, the science of measurement.

A third implication of the Josephson effect has come about by constructing a circuit containing two Josephson junctions wired in parallel. This device is known as a superconducting quantum-interference device, or SQUID, a magnetic field sensor of extraordinary sensitivity. SQUIDs are used routinely in studies of neural activity in the brain (magnetoencephalography), in microscopy, imaging and measurements of the magnetic properties of materials. A SQUID works because the magnetic field passing through the loop created by the two junctions is able to affect the interference between the superconducting wavefunctions passing along each junction, and in principle this permits a measurement of the magnetic flux to the nearest quantum of magnetic flux (the smallest unit of magnetic flux that Fritz London had posited).

Bardeen was very slow to accept Josephson's ideas and was uncharacteristically aggressive with the young Cambridge physicist, accusing him of making drastic oversimplifications with his model. Josephson patiently defended his ideas and Bardeen eventually, with delayed good grace, accepted that Josephson had been right. Josephson received his PhD, the 1973 Nobel Prize (shared with Ivar Giaever and Leo Esaki), and a chair at Cambridge, all fitting rewards for a brilliant piece of insight which has had far-reaching consequences. He has spent most of the rest of his career devoting himself to his 'mind-matter unification project' which aims to find a physical basis for extra-sensory perception, telepathy, and various other paranormal phenomena. It is perhaps unsurprising that his activities in this area have not won him the universal admiration of his scientific colleagues.

How to make a useful superconductor

Onnes had realized at a very early stage that a brilliant application for superconductors would be in wire wound around coils to make magnets. In order to achieve a large magnetic field, it is necessary to pass a large current around a coil, but the limiting factor is the heat dissipated by the resistance of the wire. With superconducting wire, the electrical resistance is zero and this problem neatly disappears. However, there is a considerable problem. Superconductivity is lost when the magnetic field exceeds a critical value and in the elemental superconductors known to Onnes the critical magnetic fields were extremely low. Even as late as the early 1950s, David Shoenberg, in his monograph about superconductivity, confessed that he thought there was little promise in using superconductors in high-field magnet applications. It was therefore vital to find materials in which the critical field could be pushed to higher values.

Niobium is a dull, soft, grey metal which was discovered in 1801 by an English chemist, Charles Hatchett, while working on a sample of the ore columbite in the British Museum in London and isolated from it the new element. He named it columbium, but it was rediscovered nearly 50 years later and called niobium, now its standard name. Niobium has the highest transition temperature of any element (9.3K) and it turns out that alloys of it give some of the most useful superconductors. In 1941, niobium carbide was discovered to be a superconductor at 16K, the record at the time, and therefore it seemed that useful superconductors might be found by choosing the right alloy. But how to choose? Although there are only a certain number of chemical elements, the number of possible chemical compounds or alloy compositions is virtually unlimited. This is because there are countless ways of combining elements in different proportions. To make progress in this area required a special type of mind.

Bernd Matthias was born in Frankfurt and studied at ETH in Zurich in the late 1930s, doing a doctorate with the Swiss physicist Paul Scherrer during the Second World War. He came to the United States in 1947 and spent the rest of his life working in various laboratories, including in Chicago, Bell Labs, Los Alamos, and San Diego. His passion was the quest to discover new materials and he pursued this quest with energy, ingenuity, and zeal. Matthias wanted to find which of the many different possible alloy compositions would provide useful superconductors. To participate in this new field of research, you need to have an instinctive knowledge of chemistry and be skilled at various techniques of chemical preparation. There were so many compositions to search through, it cannot be done at random; Matthias had to trust his instincts.

Bernd Matthias was deeply distrustful of theorists. He was a very late convert to the BCS theory (not really accepting it until many years after everyone else had) and was very aggressive about anything he thought smacked of theoretical jiggery-pokery (he was particularly antagonistic to the idea of 'organic superconductors', which were first being talked about in the 1960s [see Chapter 8 and 9],

26. Bernd Matthias

persistently making jokes about 'superconducting carrots'). His main point was that despite the most ingenious theoretical constructions obtained so far by the most brilliant minds (as detailed in the last three chapters), it was not possible for theorists to tell you whether a particular compound would be superconducting or not. They might mutter something about self-energies or Greens functions, or retarded interactions, but if you pointed to an element in the periodic table or wrote down the chemical formula of a candidate superconducting compound and asked them whether it would superconduct or not, they couldn't tell you. Worse still, they often indulged in what Matthias called 'prediction after the fact'. They could tell you why something behaved as it did *after* it had been measured but *not before*! Therefore, to Matthias' frame of mind, theoretical physicists and their fancy ideas were not worth much to the practical experimentalist and you had to make your own way in this game. This Matthias certainly did.

Matthias began working with John Hulm, a former student of Shoenberg in Cambridge, who had a lot of expertise in low-temperature physics. When they began, the record superconductor was niobium nitride (NbN), with a critical temperature of 15K, that had been discovered in Germany in 1941. By 1953, this had been pushed up to 17K with the discovery of vanadium silicide (V_3Si). In 1954, Matthias found Nb_3Sn, an alloy of niobium (Nb) and tin (Sn), which superconducted below 18K and which had a very large critical field. The transition temperature slowly crept up, until by the early 1970s it reached 23K in a niobium-germanium compound, after which progress seemed to grind to a halt.

On the way, Matthias and his colleagues discovered hundreds of new superconductors. Their strategy was to identify promising structural types and then vary the atoms within that type. One such structural type is known as the A15 structure, for reasons which are really too dull to mention, but this proved to be very fruitful; Matthias found lots of superconductors with the A15 structure. Matthias also identified a particular number of outermost (valence) electrons per atom that was favourable for superconductivity and homed in on that. His rules and tricks were based on observation and hunch, not on any grand over-arching theory, but nevertheless his cookbook approach regularly worked.

At the same time, various people tried to make magnets out of the new superconductors, discovering that wires in big coils can behave rather differently to small samples in laboratory experiments. The best conventional magnets which consume a lot of power and are filled with heavy iron cores can produce fields up to about a couple of tesla (the unit of magnetic field; the field from the Earth which causes a compass to point North is only about 0.00005 tesla). A couple of tesla was the record to beat. In 1954, George Yntema at Illinois built a magnet out of niobium wire that reached 0.3 tesla. John Hulm quickly tried the same thing and achieved double that. By 1960, it was found that Matthias' Nb_3Sn

material could give a field of nearly nine tesla. The technology to design superconducting magnets was built up and today it is possible to buy a superconducting magnet which will produce a field of over twenty tesla. Superconducting magnets had come of age.

It was just a shame that the record transition temperature had stuck at 23K and this situation persisted through the 1970s and the first half of the 1980s. Matthias' approach had probably run its course and further dramatic progress seemed unlikely. And then, in 1986, everything suddenly went crazy.

Chapter 8
High-temperature superconductivity

As the 1980s began, it seemed fairly clear that superconductivity was a pretty unexciting field with no surprising developments expected, though one could argue that surprising developments are never expected! There was an established theoretical framework in place (BCS, augmented with the insights from Ginzburg and Landau) which explained everything that had been so far discovered, and the few further discoveries that were still being made had the feeling of dotting i's and crossing t's. The record transition temperature of 23.2K (in Nb_3Ge) had stood since 1973. The BCS theory provided little hope that you would be able to find superconductors working at very much over about 20K because in this theory the transition temperature is set by various parameters, such as the energy of the lattice vibrations, and these parameters were not expected to vary greatly beyond what had already been found. It was therefore not a great surprise that a high-temperature superconductor had failed to be discovered and there was absolutely no reason to hope that such a discovery was possible.

Such pessimism had not completely drained the enthusiasm of all workers in superconductivity and a few brave souls carried on searching for superconductivity in unlikely places. In the 1970s, some unusual organic materials were synthesized by Klaus Bechgaard in Denmark which turned out to be superconducting when you subjected them to high pressure. Further chemical modification produced a material which superconducted at ambient pressure. The great thing about these so-called *organic superconductors* was that you could make small chemical changes to the molecules which comprised them and, if luck was with you, you might get a new material which still superconducted but did so differently, and possibly at higher temperature. Such research has led to the preparation of dozens of organic superconductors and some of them have been extremely important for research in fundamental superconductivity. However, the transition temperatures had not been very high.

However, the real breakthrough occurred from another, completely unexpected direction. Unexpected, that is, to everyone except perhaps two scientists working for IBM called Georg Bednorz and Alex Müller. IBM has a number of research laboratories in different parts of the world and these have led to a variety of research advances in areas of science far broader than you might expect from a computer company. In the 1980s, the IBM labs in Zurich, Switzerland, had developed a new type of microscope, the

27. J. Georg Bednorz and K. Alex Müller

scanning tunneling microscope (or STM) which works on the principle of quantum mechanical tunneling (described in the previous chapter). The STM has revolutionized microscopy, kick-started the science of nanotechnology, and won for its inventors a Nobel Prize. STM research was therefore a major and highly fashionable component of the work done at the Zurich laboratories, much more so than the deeply unfashionable work on compounds called perovskites which was carried out in the group of Alex Müller.

Bednorz and Müller

Perovskites are a type of oxide (a compound containing oxygen), found in certain minerals and named after the 19th-century Russian mineralogist Count Lev Aleksevich von Perovski. Müller was studying them because he had a hunch that because of their vibrational properties they might show interesting electrical behaviour. Müller had been energized about the possibilities of superconductivity after taking a long sabbatical in IBM's Yorktown Heights laboratory and seeing IBM's Josephson computer, a massive project aimed at utilizing the fast switching speed afforded by Josephson tunnelling devices (a project that was eventually cancelled). Although everyone else was focusing on metallic alloys or what are known as *intermetallic compounds*, Müller's attention was drawn to oxides, and in particular to perovskites.

Perovskites have the general chemical formula ABO_3, where O is oxygen and A and B are metal atoms. Their structure is shown in Figure 28: the A atoms sit at the corners of a cube, the B atoms at the centre of each cube and the oxygen atoms form an octahedron around the B atoms. The atoms A and B can take many possibilities, one of which has A=Sr and B=Ti where Sr is the symbol for strontium and Ti is the symbol for titanium; in this case the compound is strontium titanate ($SrTiO_3$), which exists either as a white powder or as a transparent crystal, the latter used in imitation diamonds for jewellery until better materials were found.

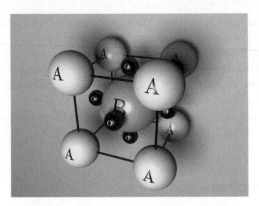

28. The perovskite structure, with chemical formula ABO_3

Strontium titanate is brittle and electrically insulating and seems
to be quite unlike the sort of material which would ever become
superconducting. However, in the late 1960s it was shown that if
you take strontium titanate and reduce it, that is remove some of
the oxygen ions, it does actually become superconducting, albeit
below at most 0.3K. This is not a result to hold the front page for.
However, it was remarkable that the effect occurs at all and was the
first time that superconductivity had been seen in, of all things, an
oxide. After all, oxides were exactly what you did not want when
you were working with metals. Iron rusts and other metals tarnish,
and all because of the formation of surface oxide layers due to the
reaction of those metals with oxygen in the air. Oxides are usually
electrical insulators and so the last things you would think of as
candidate superconductors.

A little over ten years later, a scientist at the IBM Rüshlikon
Laboratory in Zurich, Gerd Binnig, decided to see if the
superconducting transition in strontium titanate could be pushed
up a bit further. With a team which included a young physicist
called Georg Bednorz, a small amount of niobium was added to
strontium titanate in order to increase its carrier concentration
(this addition of a small amount of an extra substance is referred to

as 'doping'). The addition of niobium pushed the transition temperature up to 1.2K, a result they understood in terms of the modified lattice vibrations induced by the addition of niobium atoms. This was promising, but still seemed to be something of interest only to specialists. Gerd Binnig and his boss, Heinrich Rohrer, moved to another project, the scanning tunneling microscope (STM), and therefore were lost to the field of superconductivity; however, their invention of the STM won them the 1986 Nobel Prize in Physics, so you can hardly blame them. Binnig's young assistant, Georg Bednorz therefore joined up with another Rüshlikon scientist, Alex Müller, in a search for new superconductors based on perovskites.

This seemed a bit of a dead-end project, but some optimism was given by other discoveries in the scientific literature from the early 1970s. The perovskite lithium titanate had been reported by David Johnston, a former student of Bernd Matthias, to go superconducting when cooled below 13.7K. Another oxide, a compound of barium, lead, bismuth, and oxygen, had been found by Arthur Sleight at Du Pont to have a transition at a similar temperature. Bednorz and Müller therefore had a feeling that oxides were promising candidates for superconductivity if you could find the right ones. Since they were looking for materials in which the lattice vibrations and the electrons coupled very strongly, they started to think about materials which showed something called the *Jahn–Teller effect*, a strange instability that affects certain magnetic ions and causes the oxygen ions around them to distort. They began by looking at compounds which contained Ni^{3+}, the nickel ion with three positive charges. However, in late 1985 they spotted a report from a French group on a perovskite containing barium, lanthanum, copper and oxygen. The French scientists had found their sample to be metallic between 300°C and −100°C, but they hadn't cooled their sample down any further since they were more interested in the high-temperature properties and in particular its possible use in catalysis (as an agent to speed up certain chemical reactions).

They decided to immediately prepare oxide samples containing the same elements as in the French report but varying the ratio of barium to lanthanum in order to change the charge state of the copper. This works because barium ions have two positive charges on them, while lanthanum ions have three. Copper ions are less fussy about how many charges they have, so they easily take up the slack, so to speak. The extra positive charges on the copper ions act as 'holes' (missing electrons) and are quite mobile; they can hop around inside the crystal, thereby giving metallic behaviour. By mid-January 1986, Bednorz and Müller had found superconductivity and with some further optimization of the composition got it as high as 30K. This was a completely unprecedented temperature, but they were worried about being too hasty to announce the result. Recent years had seen a flurry of erroneous claims of high temperature superconductivity which had all turned out to be irreproducible. Therefore they were anxious to be self-critical, and in fact knew that their samples contained a mixture of two different chemical species. Furthermore, they had not been able to perform magnetic measurements in their laboratory and so could not yet demonstrate the Meissner effect. However, by April 1986 they decided it was time to submit a paper to a journal and report what they had.

The obvious place for Bednorz and Müller to publish their work was in a prestigious journal such as *Nature*, *Science*, or *Physical Review Letters*. However, they feared that the refereeing process (by which a submitted paper is sent out to independent and anonymous referees for assessment before publication) would slow down the publication of their paper. Worse, it could result in their results being duplicated by others and published elsewhere before their own work appeared in print. They therefore decided to send their paper to *Zeitschrift für Physik*. This had been a prestigious journal at the beginning of the 20th century (many of Einstein's most famous papers appeared there) but by the 1980s it had been overtaken by US journals and it eventually merged with several other European journals in the 1990s. *Zeitschrift für Physik* was not required reading for scientists working in superconductivity and therefore when Bednorz and Müller's paper appeared (reputedly after Müller persuaded the journal's editor to take it without refereeing it and to publish it immediately) hardly anyone noticed it.

Yttrium barium copper oxide

One of the few people who did notice it was Paul Chu, a physics professor in Houston, Texas. Chu's doctoral studies were performed in San Diego, where he was heavily influenced by Bernd Matthias, who instilled in Chu a determination to discover new superconductors through careful materials research and also by keeping of a watchful eye on the scientific literature for any new theory or experimental breakthrough that might help with this quest. In November 1986, Chu found himself reading an article in *Zeitschrift für Physik* entitled 'Possible High T_c Superconductivity in the Ba-La-Cu-O System' by Bednorz and Müller, and immediately realized its potential significance. He instantly set to work to see if his group could reproduce the Zurich results. Although he chose a different (and in fact faster) method of preparation, Chu's group were able to duplicate the work. Moreover, by applying pressure to the new superconductor (an experimental technique which was something of a speciality of the Houston group) the transition temperature was pushed from 30 to 40 Kelvin, setting a new record. Announcing this at a conference in early December, Chu found that Koichi Kitazawa of the University of Tokyo had also duplicated the Zurich work but in addition identified the composition of the superconducting part of their compound: the compound was now known to be the so-called '2–1–4 phase': two parts of a mixture of lanthanum (La) and barium (Ba), one part copper and four parts oxygen (or as a chemical formula: $(La,Ba)_2CuO_4$). This has the so-called layered perovskite structure, shown in Figure 29. Now physicists at other labs began to take notice. Chu knew that he had growing competition.

29. The structure of a copper-oxide superconductor. The basic unit is a copper ion (large grey ball) surrounded by six oxygen ions (black balls) in an octahedron. Layers of corner-sharing octahedra are interspersed with lanthanum (or barium) ions (small dark grey balls)

If pressure made the transition temperature rise, the next idea was to replace barium by something smaller. The element above barium in the periodic table was strontium (Sr) and using this in mid-December gave his group a superconductor with a transition temperature of 39 Kelvin without applying pressure. At the same time, an improved pressure experiment on his barium compound appeared to give him a 52.5 Kelvin transition temperature, although this proved to be rather optimistic. At the same time, similar leads were being made at various other laboratories and in particular at Bell Labs where superconductivity in a sample of the strontium compound was also observed below 36 Kelvin. The Bell

Labs group had observed the Meissner effect in their compound, providing a really convincing demonstration of intrinsic superconductivity in these systems.

Back in Houston in January 1987, Chu's team (now including a group at the University of Alabama led by Maw-Kuen Wu) began to try some lanthanum look-alikes in the periodic table in the hope of hitting upon something better. One such atom they decided to try was yttrium (chemical symbol Y). This is a silver-coloured metal and gets its unusual name from Ytterby, a village on the Swedish island of Resarö which has a large and famous quarry; the element was discovered in the quarry by the Finish chemist Johan Gadolin in 1794. Ytterby's name was also used in providing the name for three other chemical elements, ytterbium (Yb), terbium (Tb), and erbium (Er), which seems a bit excessive. Johan Gadolin did get his name attached to another element discovered in the quarry (gadolinium (Gd)), while the quarry was also the location of the discovery of holmium (named after nearby Stockholm) and thulium (named after Thule, an old name for the Nordic countries). Whether because of, or despite, its impressive Scandinavian pedigree, yttrium proved to be a smart choice. By the end of January, the Houston–Alabama team had found a new compound containing yttrium, barium, copper, and oxygen, with a superconducting transition temperature of 93 Kelvin.

Chu realized that he and his colleagues had made a key breakthrough. The new compound took the transition temperature through an important threshold: it was above 77K, the boiling temperature of liquid nitrogen. The thing about liquid nitrogen is that ordinary air is 80% nitrogen and so the raw material is inexpensive and plentiful; liquid nitrogen refrigeration is cheap and cheerful and so a material which remains superconducting at and above 77K has many technological applications. But now Chu was faced with the same dilemma that had challenged Bednorz and Müller: where to publish the results?

Chu wanted to submit his results to *Physical Review Letters* but feared that the referees would steal and replicate his results. The problem was that once you knew the formula for the new superconductor it was pretty easy to prepare it. The synthesis was relatively easy and followed what inorganic chemists have christened 'shake and bake'. You simply take your starting ingredients in the right proportions, grind them up, stick them in a temperature-controlled oven (known rather quaintly as a furnace). and bake them for a certain period and, hey presto, you have your new superconductor. Chu rang the journal editor, gave his result in outline, and requested, because of the intense competition surrounding this breakthrough, that his paper be published without review. The editor said no. Chu eventually negotiated that the paper be reviewed by referees whose names were agreed between the two of them; this was a highly unusual concession since an author normally has no say in the choice of the referees.

Chu was still not sure he could trust even his named referees and worried that the secret formula of his new compound would leak out. He therefore changed the formula of the compound in his manuscript, substituting ytterbium (chemical symbol Yb) for yttrium (chemical symbol Y), and also slightly altered the ratio of chemical constituents. He then waited until the journal said that his manuscript had been accepted, and then he waited for the manuscript proofs. At the last moment before publication occurred, he sent the journal the corrected version so that the final published version was correct.

In a delicious irony, it turned out that the fictional ytterbium compound also superconducted, though not at such a high temperature. As Chu's work was published, a number of groups 'discovered' the ytterbium compound and felt aggrieved that they had been duped, though how the 'secret' had leaked prior to publication remained a mystery. Chu came under fire for misleading the scientific community and his basic ethical principles were questioned: he had submitted a scientific paper to a major international journal with details that he knew were incorrect. Chu maintained that the incorrect chemical formula was an innocent mistake, but few believed him. The incident received considerable attention from various scientific periodicals, with one editorial writer unable to resist headlining their article with 'Yb or not Yb? That is the question.' However, when the opinions of scientists at the time were polled, most said that although Chu had technically been guilty of slightly underhand behaviour, the subsequent disclosure that his results had leaked from the refereeing process rather vindicated his tactics, and most said they would have done exactly the same.

The Woodstock of physics

The American Physical Society March Meeting is an annual event when thousands of physicists meet up to discuss the latest results in physics. The March Meeting following the breakthroughs of Bednorz, Müller, Chu, and others was one of the most extraordinary of recent times. At the superconductivity session, it was standing room only and there were so many speakers that each was granted only a few minutes to make their presentation. The session ran until the small hours and the excitement and cascade of new discoveries led to this meeting being described as the 'Woodstock of physics', a reference to the legendary rock concert of 1969. In March 1987, it was Alex Müller and Paul Chu in the roles of Jimi Hendrix and Janis Joplin, and the audience were high on physics rather than illegal substances, but the atmosphere was apparently similar.

The following few years were a period of completely frenetic activity. New superconductors were discovered every week, groups around the world abandoned their existing research programmes and rushed headlong into high temperature superconductivity, new institutes were founded, research grants flowed, and hundreds of papers were published. The new superconductors were relatively

30. The Woodstock of physics

easy to make, but were also easy to make badly, and no-one had much time to check whether the samples were of the best quality. Thus many of the early results were misleading, incorrect or both.

In July 1987, while deep in the Iran–Contra hearings, President Ronald Reagan announced an 11-point 'Superconductivity Initiative', and hailed the promise of new superconductors as 'a quantum leap in energy efficiency that would bring with it many benefits, not least among them a reduced dependence on foreign oil, a cleaner environment and a stronger national economy.'

The politicians were thus convinced that superconductivity had the potential for a technological revolution. Different nations responded in different ways. Some, like the UK, set up a single interdisciplinary research centre in Cambridge where funds were concentrated; others, like Germany, spread the money more broadly and funded interconnected university research clusters. However they did it, research funding increased.

There were several ways in which Bednorz and Müller did not fit the typical pattern that might be expected for scientists making an astounding breakthrough. For example, astounding bursts of insight are often associated with a young genius whose mind is unencumbered with decades of familiarity with the accepted views and whose time is also free of the administrative, nurturing, and leadership duties that tend to tie up senior scientists; however, Alex Müller was in his late fifties while doing his pioneering work on superconductivity. More importantly perhaps, Bednorz and Müller's breakthrough was not even a serendipitous and totally unexpected discovery since they found precisely what they were looking for. The surprise here is that they were relative outsiders to the field and were working very much against mainstream opinion. Moreover, they were working with rather limited funding on a project which had no likelihood of success; it was later commented that if their original research proposal had been submitted to a university funding agency in the early 1980s, it would have been unlikely to have received a grant. Fortunately, they were working in an industrial laboratory which at that time took a fairly open-minded view about blue-skies research.

This raises interesting questions concerning the strategy of research funding bodies in choosing which projects to invest in. The temptation is to capitalize on breakthroughs already made and pour many resources into the 'obvious' opportunities. The lesson of Bednorz and Müller is that, at least sometimes, problems are best solved when you don't attack them head on. Identifying bright people and giving them freedom to follow their instincts, however apparently unlikely to succeed, can be a highly effective strategy. On the other hand, it is obviously impossible to fund every unlikely direction in the hope that some strange and unexpected idea will turn up and so a difficult balance must be struck.

Bednorz and Müller brought physicists into the era of what is called *high-temperature superconductivity*. The record transition temperature currently stands at 138K at ambient pressure; and there is evidence that you can push this up twenty degrees or so under high pressure. This means that the term 'high temperature' is a relative one. After all, the lowest temperature recorded on Earth is 183K (−89°C) in Antarctica, so high temperature superconductivity should not evoke visions of sun-drenched beaches and palm trees. However, the reason such excitement was generated is that, for the first-time, room-temperature superconductivity seemed within reach. The BCS theory which forbade such a prospect had been blown out of the water. After 1986, nothing seemed ruled out.

Chapter 9
The making of the new superconductors

The new research field of high-temperature superconductivity
that was born in the late 1980s was radically different from
superconductivity research in the 1960s. Because 1980s research
was centred around the preparation of new oxide materials, it
required the expertise of solid-state chemists (who would prepare
the new compounds) as well as solid-state physicists (who would
measure their properties). The oxide materials were highly brittle,
and therefore material scientists were also needed to work out how
on earth to make wires out of the new compounds. The research
was therefore highly interdisciplinary and transcended the
traditional confines of physics, chemistry, or metallurgy.

Making new superconductors

The early years of high-temperature superconductivity research were
also characterized by a rapid outpouring of results on often badly
characterized materials. This is because it is relatively easy to prepare
an oxide superconductor using the 'shake and bake' technique
described in the last chapter. However, it proved to be exceedingly
difficult to make a high-quality sample, free of impurities and
defects, and with the correct oxygen stoichiometry (that is, the
correct amount of oxygen, without leaving in place some oxygen
vacancies, holes in the structure where an oxygen atom should go).
For a start, you have to know what temperature to set your furnace,

how long to heat the powder, how quickly or slowly to cool it down afterwards, and whether and how to control the atmosphere of gas inside the furnace. These choices can make or break the quality of what you produce and have to be made by people with experience and patience: the patience to repeat the whole process again in slightly different conditions to optimize the final product. Part of this process is to use various analytical tests of structure and composition to assess the quality and nature of the final product.

Many early samples were made by physicists who had no experience (and certainly no interest) in solid-state preparation techniques. They nevertheless rushed to measure and then publish the results of their shake-and-bake compounds, without taking the time to see if what they were measuring was what they thought they were. As time went on, the situation improved and techniques were developed to improve the oxygen stoichiometry problem.

New compounds steadily appeared, with bismuth-containing copper oxides pushing the transition temperature up to 110K in January 1988, and thallium-containing copper oxides nudging it up to 120K a month later. Progress then slowed a little, but some mercury-containing copper oxide compounds were found to be superconducting up to about 135K at ambient pressure in the spring of 1993, with the application of pressure pushing this above 150K later that year (see Figure 31).

All of these compounds have several features in common. They all contain layers of copper and oxygen, separated by other atoms. The superconductivity appears to be associated with electrons hopping in the copper-oxygen planes. Perhaps the most intriguing feature is that the compounds you first try and make are usually magnetic and not superconducting; it is only by chemical doping, that is by substituting one atom of one charge by one of another in the layers between the copper oxygen planes, that superconductivity is achieved. This doping seems to destroy the magnetism and introduces superconductivity. The more you add dopants, the

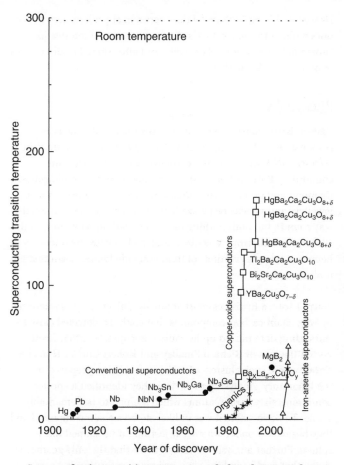

31. Superconducting transition temperature of selected superconductors since the discovery of the first superconductor

higher the transition temperature, but this only works up to a point. At higher doping levels the transition temperature starts to go down again. The highest transition temperatures are achieved with so-called 'optimal doping', but finding the optimal doping level requires an experimenter to make a lot of compounds.

However, this rapid flurry of discoveries in superconductivity was not confined to copper oxides. The renaissance of interest in superconductivity in the late 1980s and after spread to discoveries of other remarkable materials.

Buckyballs

Carbon is an amazing element and its intricate chemistry is responsible for the complexity and variety of the compounds in biology: DNA, proteins, and carbohydrates are all carbon-containing. Pure carbon exists in two main forms (or allotropes): diamond and graphite, two utterly different substances used in jewellery and pencils respectively (you have to get them the right way round), but both nothing more than carbon. However, in the 1980s a new allotrope was discovered and as with the discovery of helium, it was observations of things outside Planet Earth that got it all started.

Harry Kroto, a British chemist at Surrey University, was interested in long-chain carbon compounds that could be detected close to stars using data picked up by radio telescopes. In 1985, Kroto visited the labs of Richard Smalley and Robert Curl in Houston, Texas, where such clusters of carbon were being prepared in the laboratory. Analysing their data, they identified a species containing sixty carbon atoms which seemed to be remarkably stable. To be so stable it was unlikely to be a chain compound and they began to wonder whether it just might be shaped like a sphere. Further analysis convinced them that the only geometric shape that could combine sixty carbon atoms into some sort of spherical structure was a set of interlocking hexagons and pentagons, exactly as is found on some (soccer) footballs. Kroto remembered the geodesic dome designed by the architect R. Buckminster Fuller and so christened the new molecule *buckminsterfullerene*, though C_{60} soon began to be referred to as a 'buckyball' or a 'fullerene'.

It turned out that buckyballs are indeed formed in interstellar dust and also can be created in soot particles given off by a candle, so buckyballs have been with us for a long time. However, once chemists found a way of making C_{60} in sufficient quantities, they could begin to make compounds with it. Remarkably, when they began to do this, they discovered some new superconductors.

A profitable route was the combination of an alkali metal, such as potassium, with C_{60}, which resulted in the compound K_3C_{60}. Here the buckyballs stack in what is known as a face-centred cubic lattice (one of the arrangements a greengrocer could easily make when arranging a display of oranges) with the potassium atoms sitting in some of the spaces left between the stacked buckyballs (these spaces are called interstices), see Figure 32.

By making different substitutions, for example with larger alkali atoms, it is possible to expand the lattice, thereby causing the

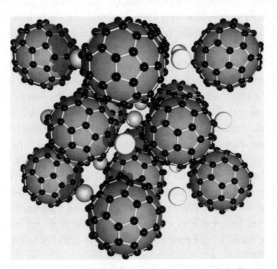

32. A fullerene superconductor composed of C_{60} buckyballs (shown as dark grey footballs) and additional ions (shown as light grey spheres)

electrons on different buckyballs to overlap less strongly. This causes the energy levels of the electrons to have a narrower distribution of allowed values and means that more electrons are able to participate in the superconductivity. This raises the transition temperature. In this way, the transition temperature has been slowly increased up to its current record of 38K in the compound Cs_3C_{60}.

Going organic

Even before C_{60} had been discovered, scientists had wondered about the possibility of organic superconductivity. Here, the word organic is not used to mean free of pesticides and insecticides. Its sense is as in 'organic chemistry', meaning the chemistry of carbon-containing compounds. As we are carbon-based lifeforms, organic compounds are highly relevant to our own biochemistry, but might some of them superconduct? Bill Little at Stanford University had speculated in 1964 that a particular type of synthetic polymer (a long-chain molecule with 'branches' that needed to be very highly polarizable) might be able to superconduct above room temperature. Little's proposal involved a mechanism called excitonic-pairing which acts between electrons and holes (absences of electrons). The mechanism turns out to involve quite a lot of energy, set by the highest energy of an electron (known as the Fermi energy), in contrast to the BCS mechanism, which involves the much smaller energy of a lattice vibration. This means that the transition temperature in Little's model would be much higher than in conventional superconductors. It was a novel idea, but molecules of the type that Little proposed seem not to be possible to make. Bernd Matthias was, as mentioned before, famously sceptical of this development, pouring scorn on a planned conference on organic superconductors in the 1970s, questioning how you could have a conference on a subject that didn't exist.

However, the necessary motivation for finding an organic superconductor now existed and various teams set out to find one.

The first breakthrough came from work by Klaus Bechgaard who prepared a series of compounds, now known as Bechgaard salts, which involve a molecule called tetramethyltetraselenafulvalene (fortunately abbreviated to TMTSF) which can donate an electron to another chemical fragment (in much the same way that sodium donates an electron to chlorine in the compound sodium chloride, which is common table salt). One of these Bechgaard salts, $(TMTSF)_2PF_6$, was found to superconduct below 1K, but only when subjected to 12,000 times atmospheric pressure. Before long, a large number of organic superconductors had been discovered with transition temperatures up to about 12K at ambient pressure. Organic chemistry is such a complicated and rich field it seems almost certain that there are many more organic superconductors waiting to be discovered.

The one that got away

At the turn of the 21st century, it was quite clear that any completely new superconductor that could turn up (and maybe even be useful) was going to be some fearsomely complicated chemical with lots of different atoms in it. After all, every element had been tried, and surely all the binary and most of the conceivable ternary compounds had been looked at. Anything that remained to be discovered was likely to be quite chemically exotic.

About the end of 2000, a rumour started circulating around the physics community that a Japanese scientist had found a new superconductor and that the discovery was exciting. Jun Akimitsu from Tokyo made his announcement at a conference in Japan in January 2001 and it took everyone by surprise, just as it had taken him and his group by surprise. It turned out that he had been trying to isolate a more complicated compound but that his sample had an impurity phase in it that seemed to be superconducting up to 39K. He isolated the impurity and found it to be magnesium diboride (MgB_2). This incredibly simple compound had been known since the 1950s and was sitting in a chemical jar in pretty much every chemistry laboratory in the world. Nobody had ever thought to measure its electrical conductivity at low temperature and so, for nearly fifty years, this extremely good superconductor had remained undiscovered. But what was more, it seemed like it was going to be useful.

A transition temperature of 39K is pretty respectable and means that with modern cooling techniques (employing what is known as closed-cycle refrigeration) it is possible to get MgB_2 wires suitably cold without needing to use liquid helium. The critical field can be as high as thirty tesla, though the field up to which you can easily pass current (what is known as the irreversibility field) is often about four times lower than that. This means that it should be no problem to produce an electromagnet of several tesla wound from MgB_2 wire. Crucially, MgB_2 is extremely cheap and when it comes to making something useful, material cost is very important.

How does the superconductivity in MgB_2 work? The first thing to notice is that magnesium and boron are both atoms near the top of the periodic table and are therefore light. This means that the lattice vibrational frequency (inversely proportional to the square root of the atomic mass) is high and, assuming that lattice vibrations are involved, this implies that superconductivity is associated with a large energy. This latter point was checked by measuring the isotope effect and it was found that the transition temperature of MgB_2 with ^{11}B was lower than that with ^{10}B, demonstrating that lattice vibrations are involved. Because pretty much every laboratory in the world had a bottle of MgB_2 sitting somewhere in a cupboard, Akimitsu's discovery was quickly replicated worldwide days after his announcement and the subsequent progress was extraordinarily rapid; the isotope effect experiment was first done in Ames Laboratory, in Iowa, and incredibly their paper was submitted by the end of January 2001, barely two weeks after Akimitsu's talk.

MgB_2 has another surprise up its sleeve. Its structure consists of hexagonal layers of magnesium interleaved with honeycomb layers of boron. The boron atoms are linked with a two-dimensional network of bonds connecting them to neighbouring boron atoms, but also have a three-dimensional network of bonds connecting them to adjacent layers. It turns out that these bonds give rise to two different species of electron which each form their own superconducting condensate and so, unusually, the superconductor has two energy gaps. This unexpected phenomenon leads to some unusual properties which have kept several physicists amused and busy, and two-gap superconductivity is now showing up in other systems.

Superconductors out in the elements

Light atoms can be helpful to superconductivity because the lattice vibrational frequencies tend to be very high and this pushes up the transition temperature. This fact has led a number of physicists to think about the lightest of all atoms: hydrogen. In fact, solid hydrogen is not a superconductor, but there is every reason to think that if it is squashed, it might be; if this could be realized, it is very likely that the temperature at which it would become superconductivity would be high. In the 1930s, it was realized that high pressure applied to hydrogen would first force it to become metallic. Even that has yet to be achieved, despite experiments which have subjected solid hydrogen to over three million times atmospheric pressure. The observation of the metallization of hydrogen remains one of the holy grails of high pressure physics.

However, there is another way to pressurize hydrogen and that is to incorporate hydrogen inside a solid. In 2008, experiments were reported on silane (SiH_4). Because of the confining effects of the arrangement of atoms in this compound, the electron density on the hydrogen atoms is equivalent to hydrogen compressed to roughly a million times atmospheric pressure. Applying even more pressure to this compound from outside drives it from an insulator

33. The periodic table. Elements that superconduct at ambient pressure are shown as black squares. Those that can be made to superconduct in special circumstances (e.g. under pressure or in thin-film form) are shown as grey squares. The temperature listed is the critical temperature, below which superconductivity occurs

118

to a metal and then, at an applied pressure of roughly a million times atmospheric pressure, to a superconductor with a transition temperature of 17K. These experiments seem to have highlighted a very promising avenue for future exploration.

High pressure can be applied to other chemical elements in order to attempt to induce superconductivity, and it frequently works. Sometimes, making the chemical element in thin-film form does the trick. The periodic table in Figure 33 shows the currently known superconducting elements: those in black are superconducting under normal (ambient) conditions; those in grey can be induced to become superconducting with a bit of brute-force persuasion. Superconductivity seems to be a much more common state than was ever previously imagined.

How to communicate your results

Following the discovery of high-temperature superconductivity, and with the field of superconductivity moving at such a frantic pace, a new problem arose. Publication delays of even a few months were becoming intolerable. However, such a delay is not uncommon in the publication of scientific articles, which can sometimes take more than a year to appear after initial submission. A submitted research paper has to be sent by the journal to one or more anonymous referees who examine the paper and provide a critique. The authors of the paper are then given an opportunity to revise their paper in the light of this critique and to respond to the criticisms raised by the referees. Sometimes, there can be a number of to-and-fro exchanges between the authors and the referees, and finally the journal editor has to make a decision whether or not to accept the paper. Once accepted, the paper must be typeset and the authors must approve the proofs. This lengthy process of evaluation and checking is not a problem during periods of 'normal science' and can often be beneficial to the quality of the finally published articles. But when a race is on and new materials are being discovered daily, issues of priority become paramount, and the race to discover is tied to the race to publish. Even though a journal keeps a record of when manuscripts first arrive in their office (and this date is printed on the final published paper), and this can help sort out disputes concerning priority of discovery, the slowness of publication means that the traditional journals are almost useless for anyone attempting to keep up with the *current* state of the field.

However, before a paper is published, 'preprint' copies of the manuscript are usually made at the time of submission to the journal and these preprints can be mailed to likely interested readers, who are more often than not competitors. Sending a preprint is a way of saying: look, we have done this work, our paper is submitted to a journal, and whatever you may subsequently claim, we got there first. In the months after Bednorz and Müller's discovery, a veritable snowstorm of preprints was being circulated between researchers. It was more than a full-time job to read these preprints and digest their contents and the people needing to do this were of course working flat out on their own research. In April 1987, in order to try and help more effectively communicate the frequent breakthroughs which were occurring, a physicist at Iowa State University, John Clem, founded a newsletter called 'High-T_c update' which attempted to digest the latest preprints and provide an intelligent commentary to put them into context. It was not long before 'High-T_c update' became an electronic newsletter (since physicists were amongst some of the earliest users of email). In 1991, Paul Ginsparg, a physicist at Los Alamos National Laboratory, set up a preprint server on which scientists could post their latest 'e-prints' and make them freely available to anyone in the world. These were unrefereed preprints, but most papers uploaded to the Los Alamos preprint server (now known as arXiv.org) were subsequently accepted by peer-reviewed journals. The papers could initially be requested by sending an email with a coded instruction in the subject line, but as the internet became established it could all be done with the click of a mouse. Although it is possible to update and correct papers after submission, the earlier versions are still accessible and their dates and times of submission listed; thus it is transparent to the research community who had which idea and when. In periods of revolutionary science, when new discoveries are being made daily, arXiv.org (now containing more than half a million scientific preprints) provides a wonderful mechanism for keeping track of these breakthroughs and sorting out claims of priority.

Magic hands

Probably the most curious affair in the history of superconductivity occurred at the turn of the 21st century in one of the world's most prestigious industrial research labs, and in fact the very place where Bardeen and Brattain had invented the transistor. It turned out to be the most elaborately conceived deception that has been perpetrated in physics.

It all started because of an intriguing possibility: by injecting charge into an insulator or a semiconductor, one might be able to push the highest available energy for carriers out of the gap and into the band, turning the insulator into a metal and possibly a superconductor. A group headed by Bertram Batlogg at Bell Labs tried this in a research programme that lasted from 1998 until 2002 and reported dramatic results: they coated an organic insulator with a thin oxide layer and fixed an electrode to the top, and with contacts on the organic insulator they had made a field effect transistor. Applying a voltage to the electrode induces a flow of charge into the organic insulator and drives it into a superconducting state. The work was pioneered by Jan Hendrik Schön, a *Wunderkind* working in Batlogg's group, who had the 'magic hands' for doing these kinds of experiments (as one fellow researcher admiringly commented) and who headed the author list on each of the papers reporting successive breakthroughs in this field. These included the discovery of both the integer and fractional quantum Hall effects (special collective properties of an electron gas which had produced two Nobel Prizes in the previous two decades), a remarkable laser effect, and also superconductivity in layers of organic compounds or buckyballs or copper-oxide superconductors. The scientific community was wowed by these results as they appeared during 2000 and 2001, with astonishing rapidity, in a succession of 15 landmark papers in the pages of the journals *Nature* and *Science* and dozens published elsewhere. On hearing Batlogg's group present the work of their group at a conference in Austria in 2000, I was convinced that Batlogg and Schön were well on their way to Stockholm to collect a Nobel Prize.

The surprising thing about the results though was that the voltages used did not cause electrical breakdown at the thin oxide layer, the effect that defeated everyone else who attempted to replicate the experiments. In fact, the problem plagued the Bell Labs team too and reportedly the effects could only be got to work on a small percentage of the devices they fabricated. In fact, Schön would need to turn up the voltage on each of his devices to see the effects, but this would result in the eventual destruction of even the successful cases, so there were never any surviving samples to analyse at the end of the experiment. But in fabricating them, there had to be some trick that Batlogg's group were using that no-one else had figured out. Nevertheless, everyone realized that his team were very smart and they were working at the best-funded laboratory in the world.

The truth proved more shocking. When it became clear that no other competitor group could get even close to replicating the results, doubts began to set in. Then some researchers at Cornell and Princeton noticed that some of the data in different papers, purportedly on different materials, looked suspiciously similar. Even the noise, the random scatter that is present to a greater or lesser extent in all real data, was reproduced in supposedly unrelated datasets! Schön quickly made a correction and apologized for a clerical error, but doubts about the veracity of the findings persisted. With further allegations swirling, Bell Labs set up a top-level committee in May 2002 to investigate the work for possible scientific misconduct. The committee reported in September of that year and found Schön guilty of scientific misconduct and he was fired from the lab (though still maintaining his innocence) and his coauthors retracted all the papers resulting from their joint work from the journals *Science* and *Nature*. It transpired that much of his data had been simulated, a fact that was difficult to establish since Schön claimed he had deleted it from his computer because of its limited hard disk space (however raw data files survived, embedded in drafts of his papers held electronically by his coauthors, and provided ample evidence of data falsification). His coauthors were found not guilty of any fraud; they had trusted Schön implicitly and had been happy to bask in the reflected glory of his discoveries and give invited lectures around the world. They probably should have been more critical in their evaluation of the work they were putting their names to, and definitely more curious about how the results were obtained. But they were taken in like everyone else and couldn't comprehend the level of deception that was being perpetrated.

The very integrity of the scientific process seemed to be under a cloud, but in fact there are some interesting conclusions to draw from the episode. First, this was probably the most audacious fraud perpetuated in the physical sciences, but it was notable for its rarity; science is founded on trust and truth-seeking and despite the rewards of scientific acclaim that might entice someone to falsify data it is remarkable that it hardly ever happens. Second, the scientific method was shown to be successful. Granted, the peer review process of major scientific journals appeared to have failed; Schön's papers were published and his results adorned the covers of the most prestigious publications. Detecting a clever, blatant fraudulent result is extremely difficult when reviewing a paper since the default action of a scientist is to believe each other to be honest reporters of what they measure. The purpose of peer review is to highlight errors of fact, method, or interpretation based on what has been reported. Nevertheless, the fact that Schön's results could not be *repeated* when tried by other groups was the trigger to increase suspicion and cause the community to scrutinize them more carefully. Within a couple of years, Schön's cover was blown.

Arsenic and old lace

The genuinely discovered new superconductors (such as the copper oxides, organics, and a family called 'heavy fermions' that I have not discussed) have an interesting feature which we have alluded to before. They seem to occur close to magnetism, in the sense that superconductivity occurs when you do something (perhaps chemically dope, or apply pressure) to a magnetic compound. Magnetism is normally a phenomenon thought of as the mortal enemy of superconductivity because magnetic fields are known to destroy superconductivity and break up Cooper pairs. This observed proximity of superconductivity to magnetism is probably highly significant and is thought to give important clues as to the origin of the superconducting pairing.

A further example of this phenomenon occurred in late 2007 when Hideo Hosono of the Tokyo Institute of Technology and colleagues announced the discovery of a new superconductor. The new compound was a pnictide. Many physicists responded with knowing nods, but secretly wondered what on earth a pnictide was and whether they had been away from school on the day that pnictides were covered. It turns out that a pnictide is a compound of a 'pnictogen', one of the atoms in a single column of the periodic table (containing nitrogen, phosphorus, and arsenic). The new superconductor had the chemical formula $LaOFeP$ and had a transition temperature of around 3K, unexciting in itself but the presence of iron (Fe) in the chemical formula was enough to raise a few eyebrows. Iron is a very magnetic atom and therefore not the sort of constituent you would expect to see in a superconductor, although in 2001 elemental iron was driven to be superconducting at temperatures of up to 2K using high pressure.

In early 2008, an analogous pnictide was prepared but this time the pnictogen was changed: phosphorous was swapped for arsenic. The transition temperature increased, and with a bit more optimization (this time replacing some of the oxygen with fluorine) a new compound La(O,F)FeAs was prepared with a transition temperature of 26K (for the correct ratio of oxygen to fluorine). Now the race was on again, and various groups tried some of the usual chemical tricks, replacing atoms with similar ones to see if that would help. By the spring of 2008, the transition temperature was up to 55K with the compound Sm(O,F)FeAs (here lanthanum=La has been replaced by samarium=Sm). A new record had been set for superconductivity without using copper (the crucial constituent of the high-temperature superconducting copper oxides discussed in the preceding chapter), with one researcher in the field enthusiastically declaiming: 'we've broken the tyranny of copper'.

These new superconductors have a layered structure, with the iron arsenic layers interleaved with the samarium-oxygen layers. It seems that superconductivity arises from a complicated pairing interaction originating from the structure of the iron-arsenic layers, although the behaviour of the electrons is rather three-dimensional. As with MgB_2, the behaviour of the electrons is rather complex and there is the possibility of more than one energy gap. Like the copper-oxide superconductors, the compounds seem to be naturally magnetic and only become superconducting upon chemical doping, with indications that the maximum transition temperature (optimal doping) occurs close to the point at which magnetism disappears. At the time of writing, it is not clear whether the transition temperature can be pushed higher or indeed what is the nature of the relationship between the superconducting and magnetic states, but these new discoveries are providing much excitement.

Somewhat less encouragingly, the presence of arsenic makes the preparation of these new compounds rather hazardous, particularly as part of the preparation involves heating a sealed tube containing arsenic and other constituents to very high temperature; if this tube explodes, you don't want to be in the vicinity.

But how does it work?

The new superconductors discovered since 1986 present a puzzle. Since the BCS theory predicted superconductivity wouldn't work much above 20K, how do these new compounds do it?

The short answer is that no-one knows. Many of the new compounds follow a mechanism that has been shown to depart sufficiently from the predictions of BCS, though the buckyball superconductors seem to follow something like a BCS model. Some people have tried to patch up BCS and find a mechanism which involves phonons; others have postulated coupling mechanisms involving magnetic fluctuations; still more have tried to consider tunnelling processes between layers. In fact, a very large number of theories are 'out there', but despite intensive effort and many heated debates over the last two decades or more, a consensus had not formed around any one.

The new superconductors have been subjected to a battery of new experimental tests and probes. The scanning tunnelling microscope has provided remarkable atomic-scale images of superconductors; various new techniques have imaged the electronic wavefunctions of the highest-energy electrons (this is known as the 'Fermi surface'), particles called neutrons and muons have mapped out the superconducting vortex lattice, and phase-sensitive experiments have revealed that the way the electrons pair in the copper oxide materials has an unusual direction-dependence (what is known as 'd-wave pairing'). But despite all this technical progress, one has Bernd Matthias' viewpoint echoing in the back of one's mind: we still understand superconductivity in real materials sufficiently poorly that we are quite unable to predict the chemical formula of an improved superconductor. Every time a new one turns up in a lab somewhere, it is very often a complete surprise. What, superconductivity in buckyballs? In copper-oxides? In pnictides? And is a room-temperature superconductor just around the corner? Before 1986, we would have said impossible. Now, most of us have a gut feeling that all we need is some clever chemistry and a bit of serendipity.

Chapter 10
What have superconductors ever done for us?

With the perspective of nearly a century of research into superconductivity, it is easy to recognize a recurring pattern. An unexpected breakthrough emerges from left field and is followed by frantic research activity. This is accompanied by feverish reporting in the scientific and popular press heralding the imminent approach of new sources of energy, new methods of transport, and other as-yet-undreamt-of opportunities for technological advance which have a bit of a sci-fi feel to them. Subsequently, the research path proves to be more uphill and rockier than first anticipated. After the initial advances have given way to much slower progress, a period of disillusionment sets in and the research grants die away. Then, after several years, the next breakthrough occurs and the cycle begins again. So what are we left with? After all this progress, what can we say superconductors have ever done for us?

Superconducting magnets

Probably the biggest use of superconductors is in making magnets. Back in the 19th century, Michael Faraday had discovered that if you pass a current down a wire, there is a magnetic field which exists around the wire. As you pass more electrical current, the magnetic field gets stronger. Winding the wire into a helical coil allows this magnetic field to be concentrated inside the coil, and hey presto you have an electromagnet, a magnet whose strength can be controlled using electrical current. Such an electromagnet is used in old-fashioned electric doorbells, and in relays and tape recorders. They are also used to make many laboratory magnets. But to produce a really large magnetic field, you need a lot of electric current, and an enormous amount of electrical power. Onnes was sure that superconductors provide the answer. However, it was only after Hulm, Matthias, Kunzler, and others in the 1960s discovered new materials with large critical magnetic fields (see Chapter 7) that superconducting magnets became a realistic possibility. The large critical magnetic fields available meant that it was going to be possible to replace the copper windings in electromagnets with superconducting wire. Although the new superconducting coils would have to be cooled with liquid helium, which is quite expensive, the current would flow with no dissipation and so the ruinous electricity costs involved with conventional magnets could be avoided. From that time onwards, various companies began to form and begin the manufacture of commercial superconducting magnets.

One such company was founded by Martin Wood, an engineer working for Nicholas Kurti in the Clarendon Laboratory. Kurti had come to Oxford in the 1930s with Simon, London, Mendelssohn, and the other exiles from Germany, and was in the mid-1950s working on studying materials at very low temperatures and in very high magnetic fields. Wood's task was to build and operate the very large magnets that Kurti needed. These big magnets were of the old-fashioned design and required enormous electrical currents to be forced through water-cooled copper coils, the electricity coming from a huge generator installed in the laboratory. Oxford Instruments, the company Wood set up with his wife Audrey, developed commercial magnets and embraced the new technology of superconducting magnets, producing equipment which could perform the experiments Kurti and others needed but with a fraction of the electrical power. The company, which had started in a garden shed, soon became one of the world's leading suppliers of superconducting magnets, and currently has an annual turnover of a few hundred million dollars.

Making a superconducting magnet is not straightforward. The metallurgical properties of superconducting alloys and the processes involved in drawing the alloys into wires have proved to be extremely complex, and variabilities in the quality of superconducting wire produced have been common. Moreover, when operating, the magnet is subject to large stresses because the action of a magnetic field on a current carrying wire results in a force on that wire; in effect, the coils are ready to burst apart due to the internal stress produced by the magnetic field. A further problem is the possibility of a so-called 'quench' that can occur if a portion of the current-carrying wire loses its superconductivity; it immediately starts to dissipate the stored energy as heat, warming the wire around it and pretty soon the entire coil has warmed up and lost its superconductivity. However, the wire is still carrying an enormous current and so the coil becomes like a kettle element, dissipating all the stored energy and boiling away the cryogens that were cooling the coil. These rapidly expand and the magnet suddenly begins to emit jets of extremely cold helium gas. Magnet quenches can be quite spectacular and so considerable effort is expended in the design of superconducting magnets to minimize their occurrence.

One of the interesting problems that needed to be solved in developing superconducting magnets was how to get the current going round the coil. A supercurrent will flow round and round the coil for ever, so how do you start it in the first place? And how do you stop it or change it? This was solved by developing a superconducting switch, a small superconducting link between the ends of the coil that can be 'opened' using a tiny heating element wound around it. When heated, the superconducting link goes into the normal state and becomes resistive; a voltage from a power source can be applied across it and the current in the magnet coil adjusted to its desired level. When this is achieved, the switch heater is turned off and the superconducting link returns to the superconducting state and current can flow across the superconducting link. The power source can now be turned off and even disconnected from the apparatus and the new current will flow around the coil indefinitely. The magnet will then produce a steady magnetic field without any energy needing to be supplied to it, though the cryogens cooling it will need to be topped up regularly.

Superconducting magnets find a variety of applications, including uses in research laboratories and magnetic separation of materials. But, as I will now describe, you will also find one in almost every hospital.

Looking inside your head

Superconducting magnets find one of their most important applications in medicine because of the development of MRI: magnetic resonance imaging. A common medical problem is that often what is wrong with a patient is located deep inside them and surgical intervention, to 'have a look', can cause more problems that it solves. What is needed is a technique which allows a doctor to have a peek inside the patient. What would be really nice is to cut the patient up in salami slices, feed the data into a computer, and then reconstruct a three-dimensional image of the bodily tissues and allow a specialist to have a really good look around. And of course it would be great if this could be done with no side effects for the patient. MRI allows you to do just this.

MRI is actually a rebranded name, because the technique was originally called 'nuclear magnetic resonance', but the medical profession cannily decided to drop the dreaded word 'nuclear'. In fact, the nucleus of every atom contains more than 99.9% of the mass of the atom, so each one of us, the food we eat, and the air we breathe, are almost entirely nuclear. But you should probably keep that one quiet; people are easily upset.

Our tissues contain large amounts of water, H_2O, and therefore lots of hydrogen atoms, and it is the hydrogen nucleus (a proton) which MRI usually focuses on. If a hydrogen nucleus is placed in a magnetic field, its magnetic moment precesses at a well-defined frequency. If an oscillating electromagnetic field is tuned to the same frequency, it can interact with the magnetic moments and energy can be absorbed. This nuclear magnetic resonance (NMR) was discovered independently by Edward Purcell and Felix Bloch in 1946, winning them the Nobel Prize six years later. NMR has found many uses in chemistry and biology, but it is its use as an imaging technique that has led to MRI. The signal tells you how much water is present and where, and it is this which gives the biological information.

The imaging is accomplished by using a magnetic field gradient (so that different parts of a patient are in slightly different magnetic fields and so the resonance occurs at different frequencies) and having the electromagnetic waves not switched on continuously, but pulsing them on and off with rather sophisticated sequences (which allows subsequent computer processing to deduce where the signal was coming from in the patient). An MRI scanner also needs a large magnet, with a homogeneous field strength. The bore of the magnet should be large enough to fit a whole patient in, or at least whatever part of the patient the doctor needs to examine. MRI scanners nearly always use a superconducting magnet. The patient is inserted into the bore of the magnet which has been charged up to a couple of tesla (roughly 40,000 times the Earth's magnetic field strength) and field gradients and radiofrequency pulses are applied. As long as the patient has not been fitted with a pacemaker or has metal implants, the procedure is non-invasive and causes no harm, though it can be a bit claustrophobic and sometimes noisy; the noise comes from the rapidly switched field gradients interacting with the main magnetic field which cause noisy expansions and contractions of the magnet coil.

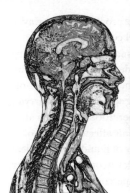
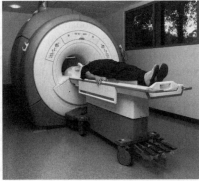

34. An MRI scan of the head and shoulders, taken with an MRI scanner (shown on the right)

MRI scans have now become routine and have revolutionized medical diagnostics. The technique can be adapted for focusing on different types of tissues and can detect tumours, examine neurological functions, and show up disorders in joints, muscles, the heart, and the blood vessels. Superconductors have given the medical profession the nearest thing to X-ray spectacles, but without the X-rays, and hundreds of thousands of people a year get a much better medical diagnosis because of them.

Particle accelerators

But it's in particle accelerators that vast quantities of superconductor are deployed. Huge superconducting magnets have been deployed at particle accelerators since the 1970s because you need large magnetic fields to bend the very energetic beams of particles that these accelerators produce. At the time of writing, the Large Hadron Collider (LHC) at CERN in Geneva has been switched on. This experiment is designed to search for the Higgs boson (see Chapter 6) by colliding opposing beams of high-energy protons inside a tunnel of 27-kilometre circumference lying under the French–Swiss border. To force the protons to travel round in this circle, a large magnetic field has to be provided all the way around the ring. Consequently, the tunnel contains 1,232 superconducting magnets, each 15 metres in length and weighing 35 tonnes. Each magnet contains coils made from superconducting NbTi cables cooled to just over one Kelvin. Nearly 100 tonnes of liquid helium are used for cooling the magnets, which are carrying over 10,000 amps of current. In September 2008, only days after the LHC was switched on, an electrical fault affected dozens of these magnets. The quench caused a tonne of liquid helium to leak into the tunnel, and the repairs led to a delay in operation of a year, emphasizing the critical role played by superconducting magnets in these large experiments.

Superconductors have also been used in constructing particle detectors for sensing X-rays, gamma rays, or exotic particles. The idea is that the particle can interact with a Cooper pair and break it up, creating excess lattice vibrations. Under certain conditions this can produce a transition from the superconducting state to the normal (non-superconducting) state which can be detected in a number of ways. Often, superconductor-insulator-superconductor tunnel junctions are used since the tunnel current is a sensitive probe of the energy distribution of electrons in the superconducting layers, which in turn is altered by the absorption of a particle. The detector is cryogenically cooled, minimizing the noise; superconductors are very convenient because unless the energy of the incoming particle exceeds the gap energy there will be little response, and so they are insensitive to background thermal radiation. In this way, various sophisticated and sensitive particle detectors can be built.

Power and levitation

Modern societies are tremendously dependent on the availability of electrical power, but people tend not to like living next to a power station. Sometimes, the source of power is tied to a natural feature such as a large waterfall or the availability of strong winds, and these natural features are not necessarily close to cities. The power must therefore be transported from power station to urban connurbation, and inevitably there are losses due to the resistance in the cables. An obvious application of superconductors is therefore in power transmission, but the need to cool the cables to very low temperatures has so far made superconductors not economically viable. The same is true in terms of using superconducting wire in transformers; there is a real benefit in terms of minimizing electrical losses, but the high refrigeration costs mitigate against widespread use.

A further problem in supplying energy is the fact that power stations tend to be on all the time but demand can fluctuate. A search for energy storage technologies is an active one and superconducting magnetic energy storage (SMES) is one of the proposed technologies. The idea is that spare energy, available at a low-demand period, is used to charge a superconducting magnet where it is stored in the magnetic field for an indefinite period; when needed, the magnet is discharged and the energy released. While this is attractive in principle, and several commercial systems are available, the high cost of refrigeration and of the superconducting coils themselves has not made this a widely deployed technique. Another energy-storage technology uses a large flywheel which is rotated at high speeds using off-peak power and thus stores the energy as rotational kinetic energy. To make this technique work, you need frictionless bearings and here superconductors can be employed. Using the effect of levitation that comes from the Meissner effect, it is possible to make non-contact frictionless bearings.

Just such a levitation effect is at the basis of magnetic levitation, or Maglev, trains. These often use conventional magnets, but the more advanced technology deploys superconducting magnets. Conventional trains are inefficient because of the frictional effect of the wheels on the track. By having the train hover over the tracks, this problem is avoided. The downside is that you need special track and this can add to the cost. The land speed record for a railed vehicle is currently held by the Japanese Maglev train which reached 581km/h (361mph) in 2003 on the Yamanashi Maglev test line. This train uses superconducting magnets to produce the levitation. A superconducting Maglev system, the Chuo Shinkansen, is planned to operate between Tokyo, Nagoya, and Osaka, although the project is at an early stage. Outside Japan, there is much less activity in developing these new train technologies: though they are potentially much more efficient than those used in existing infrastructure, the investment costs are large.

However, when you need really large magnetic fields, there is really not much choice: you have to use superconducting magnets. In building a fusion reactor, the plasma of hydrogen and deuterium is confined into a torus shape using magnetic fields and these fields are provided by superconducting magnets. If fusion can be got to work effectively on Earth (it works a treat in the Sun), then a carbon-neutral and inexpensive source of power has the potential to solve the current energy crisis. The technological challenges involved in making it work have meant that it has always been 'a few decades away' from practical realization, but this has been because it has not attracted the investment given to other major science programmes. Making fusion power a reality would be immeasurably more important than putting an astronaut on Mars. If it can be done, the power stations of the future will almost certainly rely on superconductors.

Niche applications

The examples quoted so far are all fairly large-scale engineering applications of superconductors. But there is a lot you can do on the smaller scale. Superconductors find their way into certain applications where high frequencies are needed, for example in antennas, filters, and mixers in microwave circuits, and often in tunnel or Josephson junctions. We have already described the superconducting quantum-interference device, or SQUID, in Chapter 7 and mentioned that SQUIDs are used as extraordinarily sensitive probes of magnetic fields, such as those produced by electrical activity in the brain (magnetoencephalography) or the heart (magnetocardiography).

In the early 1960s, it was realized that Josephson junctions had potential applications as fast-switching devices in digital circuits and in some cases could outperform transistors. By the end of the 1960s, Bell Labs had embarked on a research programme aimed at developing high-speed logic, and IBM began to invest millions of dollars in a project to build a Josephson computer. The Josephson junction circuits could be switched at high speed, of the order of a gigahertz, and could be miniaturized; very little dissipation was produced and the results seemed promising. However, the computer needed to be refrigerated so that the Josephson junctions would work and this increased the cost. In the early 1980s, great progress had been achieved on the project, but other technologies had moved on further, and the IBM management decided to cancel the project. Superconducting computers remain an intriguing idea but not a serious competitor technology.

Very often, many of these competing technologies lie dormant, partially developed but never quite making it in the market. This is because it is always easier for an industry to tweak an existing technology than to introduce a radically new one. But then something happens, a minor breakthrough or something dropping in price, and a new technology springs up, seemingly from nowhere, but in fact it has been there all along. A good example of this is in computer monitors. These had always been made using cathode ray tubes, even though they got better year by year, but the basic technology was the same, producing big, bulky but cheap computer monitors. Liquid crystals on the other hand used to be a slightly naff technology, only fit for using in cheap digital watches.

Then, suddenly, every new computer was shipped with a beautiful, thin, colour liquid-crystal display. Liquid crystals have been known since the end of the 19th century, but the technology only broke through rather recently.

A lot of superconducting technology is rather like that, revving up on the launch pad but never quite getting off the ground. The big disadvantage in getting more of it into the market place has been the difficulties involved with refrigeration and the fact that the newer, high-temperature superconductors have been difficult, brittle materials to work with. However, if a room-temperature superconductor is discovered, and there is every reason to think that it soon could be, everything will change.

The quantum protectorate

Science has proceeded over many years by operating a principle of reductionism, trying to understand nature by breaking things down to their tiniest constituent particles or encompassing a phenomenon by writing down a governing equation. In doing so, it has been enormously successful: Newton's law of gravity governs both the Moon's orbit and the falling apple; the properties of individual atoms can be used to explain chemical reactions and the colour of flames. And so there has been a quest into the deep fundamentals of nature in order to find the 'theory of everything', perhaps a single, beautiful equation which in itself would explain, well, everything.

The discoveries made in understanding superconductivity highlight the fact that this quest is quite misguided. In the physics of solids, we already know quite well what the 'theory of everything is': it's called the Schrödinger equation; it has been solved for small numbers of particles and found to agree perfectly and in minute detail with experiments. However, when the number of particles exceeds about ten, we rapidly run out of computing power. This is something that is not going to improve with advances in computing technology because the memory needed scales exponentially with the number of particles. In a real lump of matter, we have something like 10^{23} (that's one followed by twenty-three noughts) particles, and the computer needed to solve the equations would be literally cosmic in dimensions.

The fact that we have made progress with superconductivity in real matter has been due to the fact that it is a physical phenomenon governed by higher organizational principles, and is insensitive to some of the detailed microscopics of the problem. True things are known about superconductors, even though they cannot be deduced by direct calculation from the 'theory of everything'. The properties are said to be 'emergent' because they emerge from the higher organizational principles.

This point has been made forcefully by the physicists Robert Laughlin and David Pines who have coined the phrase: quantum protectorate. The quantum protectorate describes a stable state of matter whose generic properties are determined by a higher organizing principle, and nothing else. Superconductivity is one such quantum protectorate, while another example is the very existence of solids themselves. The idea is that the state of matter is protected by some collective quantum behaviour, so that the microscopic constituents of the system act jointly together and collectively are not affected by imperfections, impurities, and thermal jiggling around.

In superconductivity, we see the quantum-mechanical ganging up of the electrons in their many-body paired state, in which the individuality of electron pairs is sacrificed to the greater good of collective unity. Another way of putting it is that all the pairs are singing from the same hymnsheet, and this massed-choir behaviour provides protection from the detailed messiness of the microscopic system. The quantum-protectorate viewpoint claims the higher-level description to be the best one and is fundamentally anti-reductionist. After all, if by trying to understand a superconductor you break it down to the level of a single atom, you've lost your superconductor. Superconductivity cannot occur with one atom. Superconductivity only occurs in an assembly of atoms, in much the same way that an orchestral symphony can only be played by an assembly of musicians. Philip Anderson neatly summarized this viewpoint with the words 'more is different', the title of an influential article he wrote in 1972. His idea is simply that the very large number of atoms present in real bits of matter lead to fundamentally new types of behaviour which are not just simply the properties of an individual atom multiplied by the number of atoms. 'Different' behaviour occurs when you have 'more'.

This viewpoint leads to a profound realignment of the way one thinks about science. Fundamental and profound science does not have to be the science of the ultra-small particles. This is not to deny that an underlying theory of everything underpinning the microscopic physical laws really does operate. Nor is it to retreat into some kind of supernatural explanation. It is simply that the microscopic laws are largely irrelevant for a comprehensible and useful explanation of phenomena such as superconductivity. Any really complex phenomena such as superconductivity and magnetism, or even perhaps human consciousness and the sensation of free will, are emergent phenomena. They are robust against minute details concerning the microscopic laws, but are best conceptualized and described by a higher-level description. That insight encapsulates much about what is wrong with a reductionist world-view and also explains why the search for the theory explaining high-temperature superconductivity has been so challenging. Emergent phenomena require a radically new language and fresh conceptualization. Superconductivity both feeds and demands our imaginative engagement.

Dramatis personae

The numbers in square brackets refer to the chapters in which the person is featured.

- Alexei Alexeyevich Abrikosov (b. 1928), Russian physicist [6]
- Philip W. Anderson (b. 1923), American physicist [6, 7, 10]
- Thomas Andrews (1813–1885), Northern Irish chemist and physicist [2]
- John Bardeen (1908–1991), American physicist and the B of BCS [5]
- Klaus Bechgaard (b. 1945), Danish chemist [7]
- J. Georg Bednorz (b. 1950), German physicist [7]
- Gerd Binnig (b. 1947), German physicist [7]
- Felix Bloch (1905–1983), Swiss-American physicist [4]
- Louis-Paul Cailletet (1832–1913), French physicist [2]
- Hendrik Casimir (1909–2000), Dutch physicist [4]
- Henry Cavendish (1731–1810), British scientist [4]
- Paul Chu (b. 1941), Chinese scientist [7]
- Leon Cooper (b. 1930), American physicist and the C of BCS [5]
- Humphry Davy (1778–1829), British chemist [2]
- James Dewar (1842–1923), Scottish chemist and physicist [2, 3]
- Paul Drude (1863–1906), German physicist [4]
- Albert Einstein (1879–1955), German/Swiss/American physicist [4]
- Michael Faraday (1791–1867), English physicist and chemist [2]

- Richard Feynman (1918–1988), American physicist [5, 6]
- Herbert Fröhlich (1905–1991), German/British physicist [4, 5]
- Ivar Giaever (b. 1929), Norwegian physicist [7]
- Vitaly Lazarevich Ginzburg (b. 1916), Russian physicist [6]
- Cornelis Jacobus Gorter (1907–1980), Dutch physicist [4]
- Werner Heisenberg (1901–1976), German physicist [4]
- Walter Heitler (1904–1981), German physicist [4]
- Peter Higgs (b. 1929), British physicist [6]
- Pierre Janssen (1824–1907), French astronomer [2]
- Brian Josephson (b. 1940), Welsh physicist [7]
- James Prescott Joule (1818–1889), English physicist and brewer [1, 2]
- Harry Kroto (b. 1939), British chemist [8]
- Nicholas Kurti (1908–1998), Hungarian/British physicist [4, 10]
- Lev Davidovich Landau (1908–1968), Soviet physicist [6]
- Robert Laughlin (b. 1950), American physicist [10]
- Frederick Lindemann (1886–1957), English physicist [4]
- Fritz London (1900–1954), German/American physicist [4]
- Heinz London (1907–1970), German/British physicist [4]
- Joseph Norman Lockyer (1836–1920), English astronomer [2]
- Bernd Matthias (1918–1980), German/American physicist/ materials scientist [7, 8, 9]
- Augustus Matthiessen (1831–1870), British chemist and physicist [3]
- James Clerk Maxwell (1831–1879), Scottish physicist [4]
- Walter Meissner (1882–1974), German physicist [4]
- Kurt Mendelssohn (1906–1980), German/British physicist [4]
- K. Alexander Müller (b. 1927), Swiss physicist [7]
- Robert Ochsenfeld (1901–1993), German physicist [4]
- Georg Ohm (1789–1854), German physicist [4]
- Karol Olszewski (1846–1915), Austrian-born Polish scientist [2]
- Heike Kamerlingh Onnes (1853–1926), Dutch physicist, discoverer of superconductivity [3]
- Wolfgang Pauli (1900–1958), Austrian physicist [4]
- Raoul-Pierre Pictet (1846–1929), Swiss physicist [2]

- David Pines (b. 1924), American physicist [5, 10]
- A. Brian Pippard (1920–2008), British physicist [5, 7]
- William Ramsay (1852–1916), Scottish chemist [2]
- Jan Hendrik Schön (b. 1970), German physicist and fraudster [9]
- J. Robert Schrieffer (b. 1931), American physicist and the S of BCS [5]
- Erwin Schrödinger (1887–1961), Austrian physicist [4]
- William Shockley (1910–1989), American physicist [5]
- David Shoenberg (1911–2004), British physicist [6, 7]
- Lev Vasilyevich Shubnikov (1901–1937), Russian physicist [6]
- Francis Simon (1893–1956), German/British physicist [4]
- Arnold Sommerfeld (1868–1951), German physicist [4]
- Joseph John 'J. J.' Thomson (1856–1940), British physicist [4]
- William Thomson (Lord Kelvin) (1824–1907), British physicist [1, 2, 3]
- Johannes Diderik van der Waals (1837–1923), Dutch physicist [2, 3]
- Martinus van Marum (1750–1837), Dutch scientist [2]
- Carl von Linde (1842–1934), German engineer [2]
- Martin Wood (b. 1927), British engineer [10]
- Zygmunt Florenty Wróblewski (1845–1888), Polish scientist [2]

Further reading

Non-technical books and articles

P. W. Anderson, 'More is Different', *Science* 177, 393 (1972)

R. de Bruyn Ouboter, 'Heike Kamerlingh Onnes's Discovery of Superconductivity', *Scientific American* 276, 98 (1997)

P. F. Dahl, *Superconductivity: Its Historical Roots and Development from Mercury to the Ceramic Orders* (American Institute of Physics, 1997)

K. Gavroglu, *Fritz London* (Cambridge University Press, 1995)

L. Hoddeson and V. Daitch, *True Genius* (Joseph Henry Press, 2002) [on John Bardeen]

R. B. Laughlin and D. Pines, 'The Theory of Everything', *Proceedings of the National Academy of Sciences* 97, 28 (2000)

J. Polkinghorne, *Quantum Theory: A Very Short Introduction* (Oxford University Press, 2002)

More specialist accounts

A. A. Abrikosov, 'Type-II Superconductors and the Vortex Lattice', *Reviews of Modern Physics* 76, 975 (2004)

J. F. Annett, *Superconductivity, Superfluids and Condensates* (Oxford University Press, 2004)

J. G. Bednorz and K. A. Müller, 'Perovskite-type Oxides – The New Approach to High-T_c Superconductivity', *Reviews of Modern Physics* 60, 585 (1988)

P. G. de Gennes, *Superconductivity of Metals and Alloys* (Addison-Wesley, 1966)

V. L. Ginzburg, 'On Superconductivity and Superfluidity', *Reviews of Modern Physics* 76, 981 (2004)

P. Monthoux, D. Pines, and G. G. Lonzarich, 'Superconductivity without Phonons', *Nature* 450, 1177 (2007)

C. P. Poole, H. A. Farach, R. J. Creswick, and R. Prozorov, *Superconductivity* (Academic Press, 2007)

J. R. Schrieffer, *Theory of Superconductivity* (Westview Press, 1999)

D. R. Tilley and J. Tilley, *Superconductivity and Superfluidity* (Institute of Physics, 1990)

J. R. Waldram, *Superconductivity of Metals and Cuprates* (Institute of Physics, 1996)

Superconductivity

Index

Expand your collection of
VERY SHORT INTRODUCTIONS